FINDING HOME IN THE SANDY LANDS OF THE SOUTH

A NATURALIST'S JOURNEY IN FLORIDA

FRANCIS E. "JACK" PUTZ

Cypress Highlands Press of Florida
Gainesville, Florida

TABLE OF CONTENTS

FOREWORD

by Robert "Hutch" Hutchinson

"What's tomorrow like?" That's been my standard greeting to Jack for the past thirty-five years when he has returned home to Flamingo Hammock from someplace on the other side of the International Date Line, jet-lagged in days, not mere hours like less intrepid travelers. Often the cultural jet-lag between there and here is more like centuries, yet he powers through in his time machine seemingly oblivious to the century in which he is currently functioning.

Of everybody that I know, Jack is the most worldly. At any given time, he has research underway on every continent except Antarctica, with students from seemingly every country in the United Nations. He speaks a dozen Pidgin languages, often blended in stews that need no translator, in part because the humor of his language to the listener might be lost, and because translators would find the job impossibly frustrating.

It has taken somebody so well-traveled, with an intimate knowledge of the world's jungles, to describe our own North Florida in language so jocular that it camouflages the sophistication of the observations. Jack's science is a combination of 19th Century techniques applied to test 21st Century hypotheses. As a fellow who seems to be from no place or every place, he sees this place better than anyone.

Jack is easily amused. Whether he's luring fellow botanists into mudholes by insisting they get a closer look an orchid so rare it doesn't exist, or donning a hollowed-out watermelon rind as a helmet for a head-smashing duel, he has always pushed the limits of nature experiences. But a walk with Jack in the woods is also a Zen experience, as there is always something right there in the moment to see for the first time, to marvel at how natural selection has chosen beauty, and to feel the comforting solace of Mother Nature.

Jack's hundreds of scholarly articles seem to always have a few zingers deeply imbedded -- perhaps most scientists include inside jokes in their publications. For example, the word "flounder" appears far more frequently than it should, and where else might you find mention of a researcher's infant son's buttocks in a serious scientific publication about biomechanics? Despite his being an exceptionally prolific science writer, he claims to have struggled with "the voice" in this collection of essays. It seems to me that he has expertly captured his voice, a bemused observer of the ways of Floridians in their native environs.

Jack's literary legacy comes honestly -- he owes equal homage to Archie Carr and Archie & Veronica. Archie Carr lived just down the road and casts a long shadow in these woods as a scientist, teacher, conservationist, and author. Archie & Veronica are the gently joshing, G-rated comic strip characters who provided much of the moral underpinnings for baby boomers to rebel against, while we suppress our nostalgia for those simpler times with their quaint dilemmas easily solved. Jack's quirky humor mixed with incisive observations, are a meld of his Archie predecessors, and both would be proud the insights he offers.

Many contemporary crises revolve around loss of connection to place; land is the basis for our dreams, our fortunes, our loves. If you have always felt a longing to make and sustain real connections to place and people, read Jack to feel our past, savor our present, and get a glimpse of tomorrow.

Gainesville, Florida

ACKNOWLEDGMENTS

The people peppering these pages, friends and family members mostly, should share the credit or blame for this story. Many I mention are members of what, in modern parlance, could be called the "intentional community" of Flamingo Hammock Land Trust, an area of just over 100 acres that is populated by a mishmash of folks who share a love for the land. The personal histories of the original residents are varied, but most are of the Woodstock Generation and persuasion. If the twists in any of the stories are beyond the bend, Hutch is most likely at fault, especially if Zippy the Pinhead appears lurking around the corner.

Anyone who gets in trouble following my restoration recommendations, especially those that involve open flames, should note that I will have legal counsel on site, unless Richard manages to get his red tractor stuck in the process of trying to extract his jeep from the muck of Fennel Bottom or some such place. Mechanical misrepresentations can be attributed to Doug, whereas misinformation about arboreal policies can most likely be traced back to Meg, our much acclaimed but now retired City Arborist. My own nuclear family has undoubtedly led me astray on gustatory issues, although I admit that the catbrier pancake recipe needs work. That said, until my son Antonio temporarily turned into the Teenager from Hell, he was a willing subject in many of my experiments. I should also acknowledge that my daughter Juliana's whines were and remain endearing, and my wife Claudia becomes more so with every passing season.

Most of the essays in this volume are about places close to home, but many were written when I was far from Florida, so some of my acknowledgments are likewise far-flung. Perhaps my visions of home were clearer from the other side of oceans or the lines of Mason-Dixon or Wallace. Perhaps it was simply because I had more time on my hands when there weren't muscadines to prune, laurel oaks to girdle, longleaf pine seedlings to plant, and students to mentor. In any case, I started these essays in Zimbabwe, where, thanks to a grant from the European Union to the Center for International Forestry Research, I ran a research project on miombo woodland management. I thank my friends and colleagues in Harare and the village of Christon Bank for their encouragement. A post-African hiatus in essay writing ended when we spent a year in Petersham, Massachusetts, thanks to a Bullard Fellowship at Harvard Forest. Somehow the hemlocks and beavers of New England inspired me to write about the pines and pocket gophers of Florida. More essays emerged each June from 2005 through 2009, when I occupied a folding professorial chair at Utrecht University in the Netherlands, a wonderful place from which to contemplate life in Florida. I recognize that it was the mercifully short Dutch workweek that allowed me time to satisfy my professorial duties and to write, both of which gave me great pleasure. What I incorrectly believed were the penultimate edits were made in Amazonian Peru where, thanks to Claudia, I enjoyed the status to which I long aspired – the trailing spouse. Finally, several years later as this collection finally goes to print, I want to thank the generations of students at the University of Florida who listened to my stories and inspired me to put them down on paper.

The following chapters appeared previously, often in slightly modified form, in the Palmetto, which is published by the Florida Native Plant Society. I thank the long-time editor of that lovely journal, Marjorie Shropshire, for her encouragement and many suggestions on these chapters: "Seasons in the Sunshine State"; "Controlled Burns"; "Practicing and Preaching Pine Savanna"; "Restoration"; "Would You Prefer Red Coontie or White?"; "Trying to Eat Tread Softlies"; "Mockernuts"; and, "Onset of the Homogeocene, Continuing Extinctions, and Growing

Boredom." The technical articles on which I drew for some of the chapters were published in the following scientific journals: *Oecologia, Economic Botany, Ecology, Biotropica, Journal of Ecology, American Journal of Botany, Journal of Arboriculture, Forest Ecology and Management, American Midland Naturalist, Palms (formerly Principes), Conservation Biology*, and *Wild Earth*.

1. The New Floridian

Although born north of the Mason-Dixon Line, I believe that, over the past decades, I've sweat enough, shed enough blood in catbrier thickets, and donated enough of the same to various sucking invertebrates to justify calling myself a Floridian. To the extent that Florida is Southern, perhaps I qualify as a Southerner as well. Indeed, a case can be made that the farther north you go in Florida, the more Southern it becomes, and I live much closer to the Georgia line than to Miami or Mickeylandia.

Regardless of my current qualifications as a Floridian or a Southerner, I hardly fit the image of a haggard romantic or a self-destructive lunatic. Hutch, my neighbor and friend, has repeatedly pointed out that "as long as you insist on putting up firewood before the first cold snap, you ain't no Southerner." Ah well, perhaps I'll settle for being just a Floridian.

In some ways, I can justify claiming to be the quintessential Floridian. I am now over sixty, married to a Colombian, and enjoy both grits and boiled peanuts. While I am prone to dismiss theme parks and Miami-madness as not "real" Florida, I'll defend the lunacy portrayed in the novels of Carl Hiaasen, John D. MacDonald, and Tim Dorsey. That said, and given that I still shudder at the gritty hardness of Harry Crews and his kin, I have to admit to not qualifying as that sort of Floridian.

Some might say that my path to becoming a Floridian started when the plane that brought me to Gainesville landed. On that December morning in 1982, I didn't find the landscape inviting. Having recently returned from the mountains of Malaysia and the rolling hills of upstate New York, I couldn't imagine living in a place so flat, with apparently nothing but pine plantations interrupted sporadically by curved swathes of dead-looking trees. I was actually horrified to think that this monotony was to be my new home. Worse were my lingering doubts about the human inhabitants of this wretched flatness, that tasteless tribe of Southerners whom I expected to resemble some combination of the Dukes of Hazard, Scarlet O'Hara, and the residents of Yoknapatawpha County.

As the plane taxied, I thought back to those real-estate salesmen dressed in brightly colored Bermuda shorts who frequently interrupted television programs up north with invitations to "come on down." I did feel like I was coming down in lots of ways. Florida just never seemed as cool as places like Colorado, Vermont, or any and all the states in the Pacific Northwest. Most of all, Gainesville specifically did not seem like an enviable place to live – the South was scary enough, but the University of Florida (UF) seemed more like a massive football program with an associated school than a legitimate place of scholarship.

Those prejudices now haunt me in many ways. Financially, had I accepted the fact that Florida was my destiny, I would have enrolled in the State's retirement plan and would now be getting offers of hard-to-resist retirement packages. Instead, when I realized that I would not be vested in that plan unless I spent at least ten years in the Sunshine State, I opted for a transferable plan that hasn't turned out so well. But recognize that, back then, I was only twenty-nine years old, fresh out of graduate school, and – to my mind at least – destined for something other than pines in lines and neckless jocks. In my limited defense, I should say that, even then, I realized that there was a certain irony about a boy from New Jersey sneering about his new home in Florida.

My initial disdain for Florida was challenged my first morning on the UF campus as I strolled along the shore of Lake Alice. I distinctly remember that walk because I fell for the old snakebird trick. As I stood on the bank, hoping for an alligator sighting, a fish flew by, very slowly. It was what I would've identified as a sunfish, what locals might call a panfish or perhaps a stumpknocker. What struck me as odd was that it was flying very slowly a few inches above the water's surface. It took me a while to realize that it was actually an anhinga, swimming with its body underwater and holding a sunfish by the pectoral fin on its far side where I couldn't see its beak. Even after I solved the mystery, I remember thinking that Florida might at least provide some amusement while I worked my way towards a more elevated position.

My attachment to the locale was enhanced after only a few weeks in town, when I was summoned to the office of Florida Defenders of the Environment (FDE) by its president, the late Marjorie Carr. A lot has been written about that daunting woman since, but when I walked into her office, I was in no way prepared for her force field. I suppose we exchanged some pleasantries at first, but then she asked the question for which I was summoned: "What will you do for Florida Defenders of the Environment?" Her question about the organization she ran was not whether I would be willing to join FDE, but, rather, how I would contribute to it. Her assumption was that, as an ecologist engaged in understanding nature, I would obviously want to do what I could to save it. That engagement – one that she insisted on – certainly helped to make Florida my home: fighting for something gave me a sense of ownership and responsibility. Marjorie died quite a few years ago, and her absence is still felt in the form of a too-deep town-gown gap on environmental issues: too many potential contributors among the gowns who do not develop a sense of this place or otherwise heed calls like Marjorie's.

Fortunately for many of us, much of Florida isn't what it's hyped up to be. The place where I have had the pleasure of living for the last thirty-plus years is a far cry from the tropical paradises portrayed in tourist brochures, the palm-studded landscapes and beachscapes of painters, and the slick urbanism of movies and television programs. Instead, we have a half-dozen different species of pines, swamps of various profundities, a fascinating history, and a variety of ecosystems that sort themselves out on less than one-hundred feet of topography. While Lynyrd Skynyrd is still not my favorite, and I don't watch reruns of the Dukes of Hazard, I am now at home in Florida in particular and the South in general.

In the pages that follow, I attempt to describe how I found a home at about thirty degrees north of the equator at the dizzying elevation of eighty-two feet. Learning about the natural and human histories of Florida helped me find my place here; my hope is that, by describing my path, others will be more likely to appreciate what this part of the world has to offer.

Although I adore big, wild places, I focus here on the species and natural phenomena at the doorsteps of anyone lucky enough to dwell, even temporarily, in the Sunshine State and the other sandy lands of the South. From the nests of native fungus-growing ants to Spanish moss-draped live oak, one does not need to travel far from home to find wildness here. My hope is that, by exploring some readily accessible nature, I can help others develop a more fulfilling sense of this place.

2. Seasons in the Sunshine State

Winters in Florida can be brutal, especially when star-filled night skies suck the heat from the land, kumquats freeze rock-hard, and birdbaths become skating rinks. Definitely not the Florida of postcards or Tarzan movies. And isn't it ironic that, after my many years in the North, it was only after I moved to Florida that I learned that hot-water pipes freeze first?

Looking back, I am surprised how quickly after moving to Gainesville the question "don't you miss seasons?" started to bother me. Sure, Jack London would scoff at what passes for winter down here in our neck of the woods. But that said, I'd like to invite him or any of the other seasonality doubters to join me with a pipe wrench down in the crawl space under our house after a hard freeze.

I was out of town when the big Christmas holiday cold front of 1985 hit. Other than the wood stove, I had no backup heat source. To add insult to injury, I'd foolishly left no faucets dripping, and the exposed pipes were not insulated. Fine, but why do pipes always seem to burst in the worst possible places, such as behind the cabinet I had mistakenly installed in too-permanent a fashion? At that point in time, I was a bachelor, and the house that has since grown to a palatial 1,200 square feet was then only a 20 x 20 foot cabin perched on cinderblocks. The crawlspace provided a cool summer refuge for the dog, but it unfortunately also concentrated the winter's cold.

The line outside of B&H Plumbing the morning I returned and discovered the damage revealed that I was not the only one caught off guard by the hard freeze. By the time I made it to the sales counter, their stock was so depleted that, instead of the two 90-degree couplings I needed, I had to settle for four 45-degree couplings and some pipe. Ten years after that patch job, when we were expanding and replumbing the house, I recalled my intention of replacing that jury-rigged bit of work and was glad I hadn't wasted my time.

It's not like I wasn't warned about the cold front bearing down on the South. Even if I hadn't been listening to the radio, the sounds of chainsaws and splitting mauls should have clued me in that winter was on its way. All the houses in my neighborhood have woodstoves, at least for backup heat, but as recently as a few weeks prior to that fateful front, I couldn't give away firewood. I'd later learn that the fact that I was prepared for winter with a nicely stacked and seasoned face-cord of wood was perceived by my Southern neighbors as a stigma of my Yankee roots. Much later, I learned that this conclusion about my cultural limitations justified their helping themselves to seasoned wood from my stack. That tradition continues to this day, and I really don't mind, but I consequently will never know how much wood I need to make it through a winter.

The sort of Arctic air mass that blasted us that winter is referred to as a "Siberian Express" because it formed over northern Asia before marching across North America. In Florida, it flowed in after several clear, still nights followed by hoarfrost mornings. The freeze cloth covering my vegetable garden had successfully protected the plants against those initial radiative frosts but was worthless when the front's cold came with wind. The Siberian cold that froze my plumbing also killed all the vegetables in my garden but the Swiss chard.

Here in North-Central Florida, we experience only a half-dozen hardish freezes per year, but hoarfrosts are much more common. Polar and arctic cold fronts cause the really low temperatures, the sorts that blacken Spanish needle leaves and make huskies happy. These big bodies of frigid air are usually accompanied by wind, which means that freeze cloth and other approaches to frost protection generally fail. Covering low-growing plants with cloth can serve to trap a bit of earth-warmed and moistened air and may prevent frost-damage, but with the winds of frontal freezes, frost-cloth does little. Ditto for those big, air-stirring propellers you see mounted above some citrus groves. And remember smudge pots? Their heavy smoke slowed radiative heat losses to clear night skies, but were worse than worthless if there was any wind.

To show how far some people will go to protect their frost-sensitive crops, perhaps one of my own attempts will suffice. One year, I had a small patch of strawberries coming into fruit that I'd tried to protect against a nasty freeze by spraying with water all night long. This approach works because the latent heat of freezing warms the plant within the thickening ice blanket, but it takes a great deal of water. To clarify, it works if enough water is applied quickly and continuously enough, which is only possible if your well doesn't cavitate, which would result in the well pump burning out before the cold night is half over. Those were some expensive strawberries.

As a vegetable gardener, I dread these freezes, but as a naturalist and owner of an unjustifiable number of old sweaters, I relish them. On these cold days, I sometimes change sweaters several times, mostly to justify having unpacked them. As a naturalist interested in plant geography, these cold spells are intriguing. For example, pond-apples and strangler figs reach their northern limits just about where most cold fronts peter out – down around Lake Okeechobee in the center of the state – but extend 50-or-so miles farther north along the water-warmed coasts. From the other direction, species such as tulip poplar, mockernut hickory, and southern magnolia reach their southernmost limits at about the same latitude, perhaps because any farther south and they don't get enough chilling hours to flower. All these limits, especially on the north ends of species ranges, are in flux now because of rapid climate warming. Magnolias now thrive in places as far north as Philadelphia and the US Department of Agriculture regularly has to redraw its climate-zone maps.

Although mean annual temperatures are increasing all through the South, because extreme high and low temperatures are also getting more extreme, Gainesville is still likely to remain on the interface between the temperate and torrid zones. This location limits our abilities to grow many northern fruits (e.g., most varieties of apples) as well as the lush tropical species that often capture our fancies (e.g., mangos and mangosteens). But success with cold-sensitive crops also varies locally with frost pockets caused by cold-air drainage and warm spots on the southeastern shores of lakes. Winter also comes and goes with some fickleness here in the South, but with nothing like the emotionally gut-wrenching vengeance of the crocus-covering May snows of Massachusetts.

It's when the sandhill cranes start circling that we know winter is waning, although the weather judgments of those long-necked birds are sometimes untrustworthy. I still remember a warm spell one February a few years back – warm in Florida, that is, but still really nasty in Georgia. While we enjoyed the balmy weather, our friends to the north were listening to their trees break in an awesome ice storm. All afternoon long, huge numbers of cranes from Paynes Prairie circled over our house. Every once in a while, a single bird would strike out northwards, calling vociferously for his friends to follow, but to no avail. Over and over again this bird, which I assumed was a young male, tried to lure his flock toward their breeding grounds up in Ontario or Wisconsin. Finally, in the late afternoon, and boosted by a warm updraft, the flock agreed and set forth on their more-than-a-thousand-mile journey. They must have hit the cold front somewhere near Valdosta at about midnight because, at 3 A.M., we were awakened by the terrible racket of the returning birds. I don't speak crane well, but they clearly had some choice comments to make about the wisdom of males, the youth, and, particularly, their former leader.

3. A Sense of this Place

The place where I reside, and about which I am trying to convey some sense of, is inside the Kincaid Road loop in southeast Gainesville, Florida. In the official records of the County of Alachua, the property is referred to as the Flamingo Hammock Land Trust and was registered as a corporation in 1984. Flamingo Hammock was established by a small group of nature-oriented hippies who had morphed into professional environmentalists of various persuasions. The original land purchase and the diversity of livelihoods and occupations expanded substantially about a decade later with the establishment of the Woodbine Cooperative to our north. On the 300-plus acres of sandhill, swamp, hardwood hammocks, and old-fields that cover the two land trusts are houses that vary in size and style. Some are still in the process of being owner-built, even decades after ground breaking, and others were promptly and professionally constructed.

What is now our house at Flamingo Hammock grew organically from the cabin I occupied as a bachelor. When my recently widowed mother moved in, we installed indoor plumbing and then doubled the size of the dwelling by adding a first story under the existing structure. To accomplish this Building Inspection Department-baffling feat, Hutch and a bunch of our friends lifted the 20x20 foot cabin with a dozen car jacks, perched it temporarily on stacked railroad ties, built a room underneath, lowered the old structure onto the new room, and connected them. Equipped with a ladder, a hole cut in the old floor connected the two rooms. That arrangement served well even after I married Claudia and adopted her 7-year-old daughter, Juliana. A few years later, when we became pregnant and learned it was a boy, we expanded the house to a respectable 1200 square feet. The new house is complete with a second bathroom, conventional stairways, and several skylights (that I once thought were a good idea). Despite the lavishness of our abode, I have to admit that, on several occasions, visitors have made comments along the lines of "Cute place, where do you live during the week?"

I found a home in Flamingo Hammock, but for newcomers, actually finding our house takes some doing. The address typed into Mapquest will lead you not to our house but to a neighborhood in which you might feel a bit uncomfortable. Reckoning in Gainesville is usually straightforward, with streets that run north-south and avenues, places, roads, and lanes that run east-west, but that system falls apart in our neck of the woods. Basically nothing is Cartesian in Flamingo Hammock.

If, by some conveyance – balloon, airplane, satellite, broom, or Google Earth – you were to rise above where I live, you'd eventually see our house nestled down among the live oaks on the edge of what might, at first, look like a pasture. Although it was a pasture when we arrived, it is now on the restoration path back to being a longleaf pine savanna. You'd also see a few other houses, spaced out by a Trust prohibition of light trespass.

Many of the rules governing Flamingo Hammock's activities are officially recorded in incorporation documents that I am sure are registered and safely stored somewhere. Given who we were when they were drafted, these documents are amazingly elaborate. I suspect that attention to detail resulted from the fact that our founding members included Richard, an environmental lawyer, and Hutch, a challenging fellow to pigeonhole other than to say that his brain often works in peculiar ways. However they came to be, we have rules that permit free access to the entire property, encourage planting of native species but damn invasive exotics, and promote restoration of native ecosystems through controlled burns. We also have rules prohibiting interior boundary fencing, clearcutting, vehicles-of-the-Devil (e.g., ATVs, jetskis, and snowmobiles), biting dogs, and lawn flamingos, hence the land trust's name.

When we started Flamingo Hammock about thirty years ago, we didn't feel the need to regulate each other, but we were already concerned about our heirs and anyone who might buy into the land trust at some later date. The anti-flamingo ordinance, like our prohibition of velvet paintings, represented our efforts to render Flamingo Hammock unattractive to *those* sorts of people.

Flamingo Hammock isn't a commune, and we're not often all that cooperative; it's perhaps best to think of it as an unconventional neighborhood with lots of conventions. We have our share of introverts and extroverts, artists and scientists, vegetarians and bottom feeders, lawyers and rednecks (even a redneck lawyer), and university faculty and people who do real work. On both Flamingo Hammock and Woodbine there have been, and will hopefully continue to be, waves of children of the noisy, fort-building, tree-climbing, music-making, social-posing, soccer-playing, and sister-pestering varieties. To different extents, each member of the community has helped build our neighbors' houses, which vary markedly in size, style, and degree of completion. We enjoy some economies of scale but benefit even more from our diverse personalities and proclivities.

For various health reasons, every community should have a nurse, a lawyer, and a Buddhist (or something like one). To avoid slipping into grammatical lassitude, it's also handy to have at least a few former English majors. In our midst, it's also surprisingly useful to have a native Spanish speaker and wicked salsa dancer. For entertainment, we boast both the core of a rock 'n' roll band and a place that is just far enough away from the dwellings of civilized folk for them to practice. I'm particularly happy to have Doug as a neighbor because he likes computers, tractors, and other potentially labor-saving devices – he also has the mechanical acumen needed to keep them running (mostly). Hutch and Richard also like machines and have many of their own, which means that – when the gods of hydraulics and internal combustion are in harmony – mowing, log skidding, road grading, and other major excavation are carried out with the expected noise and noxious exhalations. Our neighbors, in turn, probably appreciate the enthusiasm with which I wield shovels, axes, rakes, hoes, bags of wildflower seeds, and pruning shears. I believe that everyone is also interested in, or at least amused by, the many experiments that my students and I are always conducting around the property – just scientists having fun at home. Despite our diversity of occupations and skills, one strong thread that holds us together is environmentalism, which, in some senses, serves as our shared religion.

Flamingo Hammock is in the southeast quadrant of Gainesville, where life is a bit slower and property is a substantially cheaper than in the opposite quadrant, where the houses are bigger, the faces are more uniformly white, the properties are reckoned by the square foot, and traffic jams of BMWs and Volvos are enjoyed almost daily. Southeast Gainesville has its share of problems, but if you manage to find yourself in a traffic jam, it most likely involves a lazy dog and at least one pick-up truck – feel free to drive around the obstruction, but wave to the folks as you pass by. We're actually in the county just over the Gainesville city line, which permits us to keep all sorts of poultry, hunt deer and turkeys, and drive vehicles that would stick out like sore thumbs in our mirror quadrant. Better yet, we're only a few-minutes-bicycle-ride from downtown and UF via the paved bike trail from Boulware Springs.

Although I miss hearing train whistles, every morning when I hit the rail trail on my way to work, I'm happy to have an off-road bicycle commuting option. The first half mile of that commute is the only real challenge. If I make it through the sugar sand patches and nothing from my basket is jarred out when I bounce over those exposed roots on the winding road through Flamingo Hammock, then I am up for anything. Up to Main Street, the rail trail passes through a series of natural areas. First, there's Boulware Springs, once the source of much of the city's water. North of the spring is a longleaf pine savanna restoration area. On warm afternoons, I often encounter a gopher tortoise munching cactus pads and other nasty stuff in a grassy patch that somehow avoided the laurel oak invasion. Across the bike trail is a bit of Paynes Prairie State Preserve and the more recently established Sweetwater Preserve, where a walking/biking trail loops down to the creek and crosses some really pretty seepage slopes.

Judging by the number of bomb-crater-like depressions that peppered the property when it was purchased with funds from the Alachua County Forever initiative, the Boulware Springs area is a rich place for Amerindian artifacts. Progressing northwards on the trail, I pass by the still-unmarked place where a bald eagle once dropped from an overhanging branch and nearly landed on my handlebars as it pumped its wings to get airborne. After a too-close look at its talons, I'd stopped for some heart-rate adjustment. Proceeding northwards, opposite our still-mostly-segregated graveyards – Pine Grove tucked in back and Evergreen prominently on the roadside – is where I have often encountered the marsh rabbit. At first, I mistook that brindled beast for a buffed-out cottontail, but, when I saw that it lacked a white tail, I realized that I was dealing with quite a different species. I have still never seen one swim, but hear that they do so quite readily if pursued by predators. (I wonder how they deal with alligators?) After passing the factory that was originally the home of the Alachua Tung Oil Corporation but now houses Jackson Stone Works (a maker of lovely countertops from granite imported mostly from Brazil and India), I first cross SE 4th Street and then a swamp with mostly red maples in the overstory and regal royal ferns in the understory. Just before Williston Road, I always look up to try to differentiate between the pond and slash pines – not sure that I have it down yet, but both are present.

After crossing Williston, the trail traverses the floodplain of Sweetwater Branch then climbs up to what were once the brownfields south of the Kelly Power Plant. The series of nicely landscaped ponds created by the excavation of contaminated soil seasonally hosts a variety of waterfowl, mostly coots and gallinules, but mergansers and fulvous tree ducks also make appearances. Although these ponds were created for the utilitarian purposes of removing contaminated soil and retaining stormwater, they have become a great amenity. Finally, after crossing Main Street, the trail passes by what used to be Porter's Quarters (now upgraded to Porter's Oaks). At that point, to get to work, I leave the trail, bike through the parade of coeds on sorority row, and arrive on campus.

From a broader geographical perspective, Gainesville sits smack in the center of the northern portion of peninsular Florida, with Gulf water to the west and Atlantic water to the east. During our winter, which is the preferred time for airship journeys, the sky is usually clear and the air column stable. If, by some means, we rapidly circumnavigated the planet at our latitude of about 30° N, we'd experience a lot of stable air and an incredible parade of deserts—Saharan, Arabian, Gobi, Sonoran, and Chihuahuan, in succession. The descending air mass that keeps us dry in the winter is part of the Bermuda High, which becomes the Azores High when it shifts across the Atlantic. Sounds grand, but the Bermuda High is a small and relatively weak portion of the global air circulation pattern responsible for the world's great deserts at the Horse Latitudes. When it does rain in North Florida during the winter, it's because an Arctic or polar cold front has benefited by a southward pulse of the Polar Vortex, penetrated down to our latitude, and stirred things up.

When the Bermuda High moves off the coast towards North Africa during the summer, air heated by the hot ground in the central part of peninsular Florida expands and ascends, lowering the atmospheric pressure near the ground and drawing in moisture-laden air from the Gulf and Atlantic. When these wet air masses converge, they too rise to create the paroxysm of thunder, lightning, and torrential rains that briefly cool our summer afternoons. If you have reason to fear being struck by lightning, you might want to consider living somewhere other than this, the lightning capital of North America.

Aside from the challenges of predicting the paths and strengths of hurricanes, being a weather forecaster for the Gainesville region must be pretty boring. From June through October, the forecast is usually a 60 percent chance of afternoon showers. In the winter, our weather follows that of Mississippi, Alabama, and then the Florida Panhandle. Many of the fronts that march across North America during the winter don't penetrate as far south as Gainesville, and only a few bring rain. As a consequence, winter is our dry season. The days are typically warm and the nights are cool or even cold, if the sky is clear;our mosquito populations plummet, and life is good.

Lands of our latitude and peculiar geography are scarce on the planet, and thus there aren't many places with climates quite like Gainesville's. The Yunan Peninsula is one, which helps explain why bok choy grows so well in our garden. Until global warming makes Gainesville safe for mangoes and avocados, we'll be caught between the Georgia peaches and the oranges of the other Florida. Meanwhile, we can celebrate our frost-hearty kumquats, Japanese persimmons grafted onto native persimmon rootstock, early-to-ripen blueberries, and some pretty amazing muscadines, all in their respective seasons.

My sense of this place, its seasons, and its ecological communities all continue to evolve, even after calling North Central Florida home for more than three decades. Although the explorations that follow are concentrated around Flamingo Hammock, they also extend out into the emerald necklace of natural areas that fringe Gainesville. In the chapters that follow, I try to show how reading our landscapes is enriched by an appreciation of subtle changes in elevation, a knowledge of subterranean shifts in geology, an expanded vocabulary of sand terminology, a sense of history, and a capacity to differentiate between a dozen species of oaks and half-a-dozen species of pines.

4. Florida is Not Flat

A misconception about Florida that I admittedly once harbored is that it's pancake flat. Certainly, when I first arrived in Florida, I couldn't imagine that I would find any honest-to-goodness hills. I was reminded of this mistake a few mornings back when I sat astride my son's mountain bike at the top of the hill that leads down to the sinkhole behind our house. As I contemplated the descent, my pulse quickened and I felt my chest tighten. Here, in this land of former beaches, I was confronted with topography. When we bought what is now Flamingo Hammock at a bank auction in 1984, none of us knew that the property featured such a hill.

Actually, I wasn't even present at the purchase and had no idea that Hutch and Richard were going to bid on it, on my behalf or otherwise. A group of us had been trying to buy land together for over a year, but I only heard about this particular option after the fact, when I returned from a Saturday field trip with my Ecosystems of Florida class. It was a drizzly cold day, and I was happy to be getting home to my trailer in the swamp south of Windsor.

Hutch was waiting for me on the stoop. At that time he lived in the old post office building in Windsor, so we were distant neighbors, but I was surprised to find him sitting there in the cold with a worried look on his face. "You'd better sit down" was the first thing he said. Since I was still sitting in the driver's seat of my 400-dollar, 1971 Opel, with its seats starting to leak their stuffing of Spanish moss, I recognized the suggestion as rhetorical, but I was nonetheless a bit concerned. He then pronounced, "You just bought the farm!"

Assuming that I would be willing to invest in this land purchase, at a bank auction that morning they had placed a non-refundable down payment on far more real estate than they could handle themselves. I was, of course, interested, and, as they say, the rest is history.

Our surprise hill is a pleasure in this mostly-flat landscape, but its origin is curious. In the absence of mountain-raising continental collisions or volcanic eruptions, how this hill formed certainly seemed worthy of consideration. For that matter, why Florida is elevated above sea level at all was not abundantly clear, at least not to me.

I first came to appreciate the steepness of our hill long before the bike ride I mentioned, even before we'd opened that trail. It was when, as part of an ecosystem restoration effort, Hutch and I burned off the understory vegetation around the sinkhole. We started what I will generously refer to as a "controlled burn" near the water's edge, but then, aghast, we watched as the head-high flames raced up the hill to our narrow but thankfully-sufficient firebreak at the top of the ridge. At that point in my life, I had not yet read Norman McClean's *Young Men and Fire*, nor had I studied much fire physics, but I learned quickly that fires climb hills rapidly and that the steeper the hill the faster they climb. (More on fires later because, for right now, the issue I wish to discuss is how this hill and others like it came to be in a land where the normal mountain-building processes are distant dreams.)

To understand how our hill might have formed, it's first necessary to recognize that, near the surface of the ground, the geology of our region consists of endless variations on the theme of sand over limestone. The sand layer can be thick or thin, but limestone is always water-soluble and thus becomes riddled with holes over time. It is from this perforated limestone that we draw drinking water, and into which we inject wastewater and allow lawn chemicals to leach. When the ceiling of a cave in the limestone near the surface gets thin enough, it collapses to form a sinkhole.

The sinkhole behind our house is one in a chain of similar depressions that mark the north rim of Paynes Prairie – the topographic rim, not the legal boundary of the Preserve. For the first decade of living up the hill from our biggest sinkhole, I had a hard time accepting that the steep-banked stream flowed into and not out of the sinkhole pond. In most of the world, streams flow out of ponds, not into them. But our stream starts up in the mixed black gum and cypress swamp on the Woodbine property to the north of us, crosses SE 27th Street, flows into Calf Pond, and re-crosses the road before our sinkhole finally swallows it. I like to think that months later the water in our sinkhole bubbles up out of the aquifer in Mickeylandia (taste THIS, you dirty rat!), but I am really not sure where it ends up.

Our sinkhole pond ranges from a deep puddle to a 10-acre lake. Major flooding events occur every few years, when heavy rains fail to percolate fast enough into the underlying limestone aquifer. When the waters rise, the submerged woodland paths make great canoe trails, and I enjoy torturing the floating masses of fire ants that bob around in the floodwaters. By holding onto one another, with the queen or queens in the middle, baseball-sized colonies of these little monsters can float until moored on the shore when the waters recede. Flood time should be a good time for fire ant control, but my various attempts at the destruction of these floating colonies have so far proven impractical or downright dangerous.

While sinkholes can form anywhere that's underlain by limestone, some places are more sinkhole-prone than others. Living, as we do, on the edge of a large prairie composed of a complex of coalesced sinkholes, the likelihood of a sinkhole swallowing our house is high enough that I am sure to keep our sinkhole insurance premiums paid. Less well known is the apparent relationship between sinkholes and Buicks.

After I read about a sinkhole that swallowed three cars in a busy intersection, I became intrigued by what seemed like a highly improbable series of events: all three cars were Buicks. I started tracking sinkhole stories more closely, and, after only three years, I was on my way towards statistical significance and a major discovery: Buicks were apparently being targeted by sinkholes – preferentially swallowed, as it were. Alas, before I could report my findings to the scientific community and secure my professional reputation, a BMW dealership went under and I abandoned the study. The dissolving away of our limestone foundation accounts for most of the little topography that we have in peninsular Florid; the rest of our hills are just dunes. *Karstic* is what our landscape is called, but there aren't any craggy peaks for Chinese painters here. Instead of flowing water carving away at the surface, our flows are below ground. The rate at which Florida is being dissolved away is staggering; something like a million cubic yards of limestone flows out to sea every year from our springs alone. While we might bemoan the loss of our land, quite the opposite is the apparent result.

Based on estimations of how much of Floridian limestone is dissolved away every year, geologists proposed an explanation for how, in the absence of continental collisions or volcanic activities, peninsular Florida has reached the dizzying altitudes of over one hundred feet above sea level. Apparently, in response to the lessening of the overburden of heavy rock, the lower layers of limestone actually rebound, which elevates the land surface. Sometimes, cracks form as the land rises, just as they do on the crust of bread rising in a warm oven. A long crack of this sort might explain the east-west trending series of sinkholes that includes the one behind our house. According to this seemingly reasonable explanation, the Ocala Arch is the result of solution chemistry, not deposition. A similar phenomenon accounts for why Chicago is also rising, but there, the ground's rebound results from the melting away of the mile-thick sheet of continental ice that covered it until about 12,000 years ago.

Back in Florida, under the limestone that is being dissolved away is more limestone. Go even deeper in Florida and you'll find – you guessed it – even more limestone. All that rock is the accumulated remains of millions of years of growth, death, and compaction of coralline algae, bryozoans, bivalves, and true corals. While alive, these creatures all happily made limestone exoskeletons in the warm waters enriched by nutrients from the erosion of the once-young and Himalayas-sized Smoky Mountains. If you persist in digging, as few have before, you'll eventually reach what geologists refer to, with reason, as the "basement complex."

Learning about Florida's ancient history helped me to find home here, but I must admit that, of all the really bad lectures that I have given, my accounts of the geology of Florida are among my worst. Tony Randazzo and Doug Jones edited a great book on this topic entitled *The Geology of Florida*; either of them could undoubtedly dazzle us with descriptions of limestone and basement materials, but, somehow, by the time I get down to the Miocene, most of my students need neck braces to lessen the risk of whiplash.

To make Florida's long Paleozoic story short, under more than a mile of limestone is African rock. Somehow, when the contents split apart 250 million years ago, North America took some of the sea floor previously attached to West Africa. Since then, we've been piling limestone and, much later, sand on top of these African schists, gneises, granites, and rhyolites. The masses of Precambrian rocks that make up the basement complex on which Florida grew have been twisted and shifted a great deal over the past couple hundred million years. For reasons of deep history, they do not sit comfortably next to one another, which is why we have earthquakes in Florida. Before you get alarmed, the earthquakes are mostly small, well buffered as they are by the thousands of feet of porous limestone and sand between our basement complex and the surface where we dwell.

I like to think about the sand underfoot being Smoky Mountains in origin. I also think we should recognize the contributions of the billions of little creatures whose dead bodies turned into the limestone on which that sand was very recently deposited. Going deeper, I also like the idea of standing on Africa, without jet lag and without needing a visa or inoculations. It's also a bit dizzying that every day, even as the waters wash away the limestone under my feet, I get higher. But, mostly, I'm happy with my hill, however it formed.

5. Real Rocks and Names for Sand

To see a World in a Grain of Sand
and a Heaven in a Wild Flower,
Hold Infinity in the palm of your hand
and Eternity in an hour

William Blake

The lack of rocks, even pebbles, is one of the conditions with which Floridians must contend. Hutch once opined that the lack of stones appropriate for chucking at often-undeserving targets leads Southern youth down the path to perdition. He offered this theory just before admitting responsibility for the introduction of air potatoes to Windsor, where we both resided before Flamingo Hammock. I rather doubt both claims, but aerial tubers from the Devil's potato do fit nicely in the palm, and they are good to throw.

When I first moved to Florida, I was blissfully unaware of its rocklessness. I had planned to continue a family tradition of cooking outdoors, for which I needed rocks to make a fire ring. At that time, I lived in a dilapidated trailer on the margin of the cypress swamp on the eastern shore of Lake Pithlachocco, which, at that time, I referred to as Newnans Lake. As trailers go, mine wasn't much to look at, but the setting was exquisite. It was nestled under live oaks with stag-headed cypress in the backyard and maturing slash pine in the front. I should also say that access to my home sometimes involved paddling – or at least sloshing – down a long sandy driveway. Once, I came home to find my neat stack of firewood washed away into the swamp. These slight inconveniences were more than compensated for by the occasional alligator dozing in the yard; the bald eagles calling overhead; and the choirs of spring peepers in January, tree frogs in March, and pig frogs in April.

Important to my eventual acceptance of Florida as home was the location of that swamp-bound trailer just outside the little town of Windsor; though only just ten miles out of the university town of Gainesville, my trailer felt a world apart. Windsor is sparsely populated by a distinctly Southern mélange of Baptists, hippies, wildlife biologists, and rednecks. My 60-hour workweeks and the nights I slept on a camping mat in my lab at the university didn't provide too much time for enjoying the local culture, but that changed a bit after I won first prize in the Ugliest Zucchini Contest at the first Windsor Zucchini Festival.

Like many small towns in the South, Windsor needed some sort of annual event to celebrate itself and to raise money for the volunteer fire department. Hutch was an active member of that fire brigade, and it somehow fell to him to come up with a fund-raising festival theme. Unfortunately, towns around us had already claimed mullet, watermelons, and peanuts, which made the selection of Windsor's emblem challenging. Back then, before commercial blueberry operations got going, few crops other than pine trees were profitably grown in the flatwoods that dominate the Windsor area.

I was not present at the bacchanal when zucchini was selected as the theme of Windsor's festival, but an observation about Hutch might shed light on that decision. He and I are close friends, and, although we are similar in many ways, in his formative years he was apparently inspired by the Zippy the Pinhead comics while I was more influenced by Mister Natural and Robert Crumb's less savory characters. That contrast will most likely be meaningless to anyone who grew up before or after the 60s, but Hutch and Zippy are both challenging characters.

Zucchini is a funny fruit, though it's often treated as a vegetable. In upstate New York, where I lived before moving to Gainesville, the only time locals locked their doors was during zucchini harvest season. They did this so as to keep their neighbors from dropping off their extra zucchinis. I partially remember one wine-soaked afternoon as a group of us sat around contemplating what could be done with the truck-load of "zukes" we'd harvested. I do remember Willy Kimler, now probably a distinguished history professor, wobbling off on his bicycle with a hat fashioned from a particularly large zucchini. The irony of zucchini was evident in Windsor's Zucchini Festival plans, which included the Ugliest Zucchini Contest that I might have mentioned winning. The tongue-in-cheek element is also evident in Julie Robitaille's *Windsor Zucchini Cookbook*, which includes a recipe for compost.

I prepared for the Zucchini Festival by planting several hills of that divine cucurbit in a clearing along the road to my trailer. Unfortunately, along with my woodpile, my garden floated away before the harvest, so I was zuke-less. Fortunately, the morning of the festival I had my wits about me while I poked around in the cypress swamp. When I stumbled across an old leather boot in the mud, I was inspired. It was like angels advised me to take that boot, attach some zucchini leaves, and enter it in the Ugliest Zucchini Contest.

On the afternoon of the First Annual Zucchini Festival, I sat in an audience of more than fifty souls as the panel of judges went down the row of ugly zucchinis, each numbered and assembled on a long table on a dais at the front of the main bay of the fire department building. Some of the entries really were pretty ugly – cancerous looking outgrowths and warts and discoloration abounded. When the judges came to my entry, one of them made the mistake of stating, quite clearly, "This is a boot." I was ready for just such a comment and immediately stood and announced in my most professorial voice, "My name is Dr. Jack Putz and I am a professor of botany at the University of Florida. If you judges are incapable of recognizing the years of genetic engineering that went into the development of a zucchini that looks, smells, and tastes like an old boot, then I must withdraw my entry." Of course I won, which was duly announced below the fold in the local paper.

Later that week, when I saw Bill Stern, the chair of my department, he congratulated me on my award. He then requested that, in the future, I would please refrain from identifying myself as a member of the Department of Botany. My concerns about whether he was serious disappeared at the Second Annual Windsor Zucchini Festival, when he won a prize for the best zucchini poem in Yiddish. I should add that my zucchini poem, written in Indonesian, also took first prize in that category.

For anyone who has more than just passed through, it is clear that Florida is a cultural mish-mash. Statewide, the north is more Southern than the south. Locally, city centers and new suburban developments tend to be more Northern than their rural fringes. In Gainesville (as in many cities in the northern hemisphere), it's poorer and blacker on the east side of town, perhaps as a legacy of rich folks not wanting to be downwind from Hogtown. Where I live, in Southeast Gainesville, the South is very much with us. Initially, that meant I sometimes couldn't understand either the words or the intentions of some of my neighbors. That was fine with me, insofar as moving to the South was shaping up to be something of a cultural adventure.

When I lived in that trailer out in Windsor, my nearest neighbor was Mark Brown, a Floridian and a wetland ecologist at UF. Mark quickly became, and has since remained, a close friend, but, at first, he struck me as more of a Maine woodsman than a Florida cracker. Not long on words is Mark.

On the Saturday morning in January when I moved my meager possessions into the trailer, the air was chilly – a nice day for a fire. I went out early looking for rocks to make a simple fireplace in the yard. At no time in my life would such a quest have taken more than a few minutes, but I was still rockless an hour later when I met Mark and Ted, his golden retriever.

"Morning Mark, how're doing?" I queried while patting Ted's massive head.

"Oh, not bad, how 'bout yourself?" he replied.

"Not bad, but I'm having trouble finding rocks for a fireplace. Any suggestions about where to look?" I asked.

Mark looked at me, gazed off into the saw palmettos and slash pines, and then pointed northwards.

"About how far would you say?" I responded to his gesture.

"Oh, I'd say about 200 miles," he offered.

He was right, at least on the surface. The closest rocks of the sort I needed were sitting on the hillsides of the Smoky Mountains, a long way from my new home. The other option, which is perhaps not worth mentioning, was to drill down through about five feet of sand and then a mile or two of limestone to get down to the granites of the basement complex on which Florida sits. Sure, limestone is in some ways legitimate rock, but it does not serve well for fireplaces. When limestone is heated, it turns to powder, which is how cement is made.

There are people in Gainesville who resort to buying real rocks for landscaping and other purposes, but my doing so would cause my ancestors to turn over in their cold, stony graves. Fortunately, my problem was solved a few days later when I paddled Mark's aluminum gnu under a railroad trestle that spans the Sante Fe River. There, lining both banks, was enough Georgia granite to nearly sink my ship. It was more than enough for a fire ring.

I lived in that trailer for nearly two years, but, even when I moved, I took those rocks with me. While contemplating the fires they are able to contain, I sometimes thank the railroad company for going to the trouble of bringing them to Florida. More often, I think of them as sitting on top of the sand that they will eventually become. Actually, it was rocks just like them that gave rise to all the sand in Florida.

I'll leave the poetry to William Blake, but it's still fun to pick up a sand grain and imagine its story. The one I have right here tells of a crushing encounter between the North American and African continental plates, way back in the Ordovician, some 450 million years ago. That collision resulted in the formation of the supercontinent of Pangea and, in the process, elevated the Smoky Mountains along with the rock from which this sand grain was derived. When the two plates started to drift apart about a hundred million years later, part of what started out as Africa ended up on North America. It is on that African basement that Florida developed.

Florida's sand is mostly the remains of those once-mighty Smoky Mountains, which reached as high as the Himalayas. For an intriguing reason, if we wanted to check that claim by taking all our sand and piling it back up, we would end up with an overestimate of the heights of those mountains. The reason for this overestimate is that some of our sand is derived from the now dispersed microcontinent of Avalonia, which once sat between Europe and North America.

Since the time of the Paleozoic continental collision, North America and Europe have been drifting apart, and the Smoky Mountains have been eroded down to what my Swiss in-laws refer to as hills. The products of that erosion are rocks on the slopes, clays on the piedmont, and sands on the outer coastal plain. Most of the bits of the Smoky Mountains that made it to north Florida are sand-sized particles of almost pure quartz, plus a few rounded pebbles carried down in storms and by strong long-shore currents. The smaller particles – clay and the like – mostly wash out to sea, sink to the bottom, and wait to become rocks again.

We have mostly quartz sand in Florida because quartz is a mineral that is light enough to be transported several hundred miles by water, heavy enough to avoid being washed out to sea and, most importantly, tough enough to survive the trip. Technically, the hardness of quartz crystals is derived from a hexagonal formation of six-sided prisms terminating in a six-sided pyramid of pure silicon dioxide. The boulders of granite, schist, and gneiss from which our sand was derived were left behind on some hillside in north Georgia.

I'm careful to say "quartz" sand because "sand" technically refers to particles of any material that range from those just visible to the naked eye to particles about the size of a head of a pin. Even quartz sand isn't tough enough to make it all the way down the Florida peninsula. The beach "sand" south of Boca Raton or Sarasota is mostly small pieces of coral, bits of bivalves, and chunks of calcareous algae, not Smoky Mountain fragments. When the marine sediments that became our soil were deposited in shallow estuarine waters some 5-20 million years ago, they probably included fragments of oysters and the like, but the calcium in those shells is soluble in water and has long since been leached away.

As any psammophile (Greek "psamm" means sand and "phile" means lover) can tell you, sand varies in chemical composition, grain size, sphericity, and angularity. The sands in demand by members of the International Sand Collectors Society are green, black, fluted, or otherwise weirdly shaped. In contrast, although our sand grains vary in size, most are just roughly spheroidal bits of quartz. On the ancient dunes along the central spine of our peninsula, you can find some small grains that have been rounded by being bashed about by the wind, but there's still nothing very exciting about that from a collector's perspective.

The Bedouin have as many names for sand as Eskimos do for snow, but, although we live in a sandy land, our sand-related vocabulary is relatively impoverished. Basically, locals discriminate between white sugar sand and yellow sand. If you look at some sand under a microscope, you'll see that each grain is scratched, like pieces of glass washed up on the beach after spending years tumbling in the surf. Despite their surface etchings, most of the grains of white sand are completely clear. In contrast, the pits and hollows on the grain surfaces of yellow sand are coated with a thin layer of (mostly) compounds of iron. Presumably, the white sand started out yellow but had its coatings stripped off by acids leached down from pine needles and other organic matter on the soil surface. This discrimination between white and yellow sand is important because that thin layer of iron oxide confers a slightly elevated capacity to retain nutrients and water.

While chemical differences between white and yellow sand are measurable, the limited extent to which our soil holds onto water and nutrients is overwhelmingly due to the organic matter it contains. Managing soils for crop production in Florida thereby boils down to either prohibiting the soil's organic matter from cooking out or replacing it after it's lost. The industrial option is to use the soil to hold up the plants and add all nutrients hydroponically, an approach that will remain financially viable until people start to concern themselves more about the quality of our groundwater. In our vegetable garden, we use a gentler touch that involves huge inputs of composted leaf litter, of which great quantities are yielded from sweeping our yard down to the bare sand.

6. Sixty Percent Chance of Afternoon Showers

It must be boring being a weather forecaster in Gainesville once the summer storms commence. Every day, it's the same 60 percent chance of afternoon thundershowers with highs in the 90s (this latter prediction refers to both temperature and humidity). Our summer rains often derive from the convergence of moisture-laden air masses from the Gulf of Mexico and the Atlantic Ocean. I bike to and from work, but I don't like cycling in the rain, so, from the early afternoon on, I keep my weather eye peeled for cloud formation. If I calculate it right, I can be home in time to watch the light show from the porch; otherwise, I get stuck at the university later than seems proper for an already-tenured faculty member.

As I mentioned above, if you hold your finger at Gainesville's latitude of about 30° North and spin a globe, your digit will pass over some of the world's great deserts – the same pattern occurs south of the equator at about the same latitude. These deserts are dry principally because they occur at the latitude at which the air that was heated near the equator first rose to the stratosphere, then – when it couldn't rise any more – split and moved both northwards and southwards, and finally cooled as it moved until it was dense enough to sink back down to the surface, often between 30 and 35° on both sides of the equator. Descending air masses warm as they fall, and, because their water-holding capacity increases with temperature and attendant expansion, the subsiding air dries insofar as it becomes less saturated with water vapor. Because the descending air stabilizes the atmosphere, winds are light and there is no chance for the rains that save us from baking on many a summer afternoon. This global pattern of air circulation is called a Hadley Cell (after William Hadley, an English lawyer and amateur meteorologist); we get to experience life in the Sahara when our portion of the Hadley Cell, which we call the Bermuda High, is on our side of the Atlantic. When the Bermuda High shifts across the Atlantic towards North Africa and becomes the Azores High, summer rains can start in Florida.

Although the sun never gets completely overhead where we live, at seven degrees north of the Tropic of Cancer, it certainly beats down like the dickens in the summer. With all that sun, the ground is soon sizzling. The ground heats the air above it, which causes that air to expand and rise, thereby inviting replacement stocks from both coasts. At night, as the land surface cools more rapidly than the adjacent seas, the wind direction changes. Those gentle offshore morning breezes give waves the form favored by bleary-eyed surfers. But it's the afternoon convergence of sea-breezes that cause the pyrotechnic shows for which Florida – the Lightning Capital of North America – is justly famous.

When the Gulf and Atlantic-originated air masses meet over Gainesville in the mid-afternoon, they push each other upwards. Additional lift comes from having their bottoms toasted by the hot ground. As the moisture-laden air mass rises, it also cools to the point that raindrops form. All that rising and jostling of condensing rain droplets and hailstones causes a buildup of electrostatic charges in the clouds; positive charges typically accumulate towards the top while the negative charges converge on the bottom. Those low-hanging negative charges push away from like-signed charges on the surface of the ground, which causes the ground to develop a positive charge. When the electrostatic differential is finally too much for the cloud to bear, there is a discharge, either as sheet lightning that passes from cloud to cloud or as cloud-to-ground bolts.

Americans are about as likely to be struck by lightning as they are to win the lottery, but since lots of people buy lottery tickets and about ten Floridians are killed by lightning strikes each year, it seems worth knowing something about this looming threat. Based on my research, our dog is behaving rationally when she hides under things during cloud-to-earth lightning strikes in the neighborhood, which happen about 10,000-20,000 times per year (assuming that our neighborhood encompasses a fifty-mile radius). Imagine a lightning channel 2-10 miles long delivering one hundred million to one billion volts at temperatures of 50,000 degrees. Next time the thunder peals, I may just try to make some room for myself under the bed with the dog.

7. Our Neatly Swept Yard

One way I've come to consider North Florida my home is by adopting some local customs. One custom that I really appreciate is keeping the area around the house free of vegetable matter. This swept-yard tradition, which was once widespread across the South and is still in vogue in rural communities in developing countries, is sadly disappearing at great cost, both financial and environmental.

As I write these words, our freshly-swept yard looks really fine. Before it got hot this morning, I used a rake – though not of the traditional bundle of dogwood or gallberry branches variety – to sweep around the house; it is, once again, cleared down to bare sand. I admit to having left a few sprigs of grass here and there, but not enough to bother a Buddhist or to suggest sloth (or worse) to judgmental neighbors. Although the clean sweep extends only about thirty feet from our mostly wooden house, the band is wide enough to serve our purposes.

One of the advantages of a neatly swept yard is that, from its edge, I can safely ignite the controlled burns we use to keep the flammables at bay and otherwise manage our ecosystem. In addition to being the traditional and still-best approach to "fire-wise landscaping," yard sweeping reduces mosquito, tick, and chigger populations near the house. For those who worry about snakes, a clean sweep diminishes that over-emphasized risk as well. Finally, for those of us with refined tastes in landscapes, there is also an aesthetic to a cleanly swept yard; it's not one that is shared by many people nowadays in the USA, though it's still appreciated by populations with deep roots both in the American South and elsewhere in the world.

I didn't grow up with swept yards. Admittedly, our back yard in New Jersey was always a marvelous mess of packed dirt, vegetable gardens, dog craters, woodpiles, and projects both underway and abandoned. In contrast, the front yard of our suburban home approached the American dream of a flawless sward of turfgrass from driveway to hedge.

Despite repeatedly claiming that he had sweated enough in India and Burma during the War to last two lifetimes, Saturday mornings in the summer would find my father struggling to tame the turf with our push reel mower. When gasoline power mowers dropped in price in the late 50s, we bought a used one that was cantankerous but did the job. To my dismay, my father forbade me from using ours until I turned ten. That restriction cost him a season of mowing, but then I used it gleefully until I was fifteen, several years after my peers had started charging for their efforts.

Keeping our lawn mowed was one thing, but securing a weed-free and lush sward was quite another. Remember that those were the days before the military-industrial complex successfully collaborated with lawn care businesses and golfing associations to use the technical support from turfgrass departments in major land grant universities to show suburban America the way to the perfect lawn. The result is that Americans spend an estimated thirty billion dollars per year tending thirty-one million acres of lawns. In their book *Redesigning the American Lawn: A Search for Environmental Harmony*, Herb Bormann and associates report that, in the process of trying to keep our exotic lawn grasses happy, we apply thirty-three million pounds of herbicides and seventy million pounds of pesticides, not to mention approximately 10 percent of all the fertilizers used in the USA. To arrest natural succession on our lawns we also burn six hundred million gallons of gasoline. Who knows how much of our diminishing supply of potable water is poured onto lawns, only to leach this cocktail of nastiness down to our increasingly-beleaguered reservoirs and aquifers. Conspiracy or not, these figures are distressing.

In *The Lawn: A History of an American Obsession*, Virginia S. Jenkins explains that the appeal of lawns is not new, nor is it entirely American, but the chem-lawn look is certainly a recent domestic problem. Sure, lawns abound in 18th and 19th Century English and French landscape paintings, but careful scrutiny reveals that the lawns portrayed were actually sheep-grazed meadows where dandelions were a pleasure. I grant that a well-kept lawn will cushion a fall and I even admit to liking the smell of freshly-mown grass, but at what cost do I appreciate these pleasures?

Adding our sandy soils, high temperatures, and heavy rains to the fact that we drink groundwater, Florida is a particularly bad place for lawns. That all our common turfgrasses are exotic species is one indication that lawns were simply not meant to be in our ecosystems. Given the low capacity of our soils to retain nutrients, highly soluble inorganic fertilizers applied to lawns quickly leach down beneath the reach of grass roots and then sink down further into the aquifer. From there, they emerge in our springs or end up in our drinking-water supply. High nitrate concentrations in both our ground water and springs are a testimony to the widespread and inappropriate use of fertilizers; far too much of that misuse involves turfgrass lawns.

Feeding a lawn with nitrogen greens it up, but, because grasses, unlike trees, do not continue to accumulate nitrogen-containing biomass, excess nitrogen is quickly lost to the system. What doesn't leach downwards is broken down by microbes into a variety of gaseous nitrogen-containing compounds, many of which atmospheric chemists refer to as NOXs. These compounds are of concern because they contribute to both smog formation and global warming. Add the NOX problem to the petroleum used in the Haber-Bosch process to "fix" gaseous nitrogen into fertilizer, and you can see why concerned citizens opt for small lawns and organic fertilizers.

It is clear that, for the health of both the atmosphere and our aquifer, much of our lawn acreage should be replaced by swept yards or other more environmentally-friendly landscape alternatives. Books like *How To Get Your Lawn Off Drugs* provide plenty of advice, such as switching to organic fertilizers instead of the 18-8-8 that you can buy at any garden or home improvement store. Some consumers are put off by the higher price per bag of organic fertilizer, not realizing that, because the nutrients are released slowly, less fertilizer is needed and the cost of maintaining a green lawn is therefore actually reduced.

I was surprised to learn that the high carbon cost of modern chem-lawns is not due to the burning of fossil fuels in mowers and edgers or the natural gas used to industrially "fix" all that nitrogen. Instead, the major "carbon cost" of modern turfgrass lawns in the South is the fuel used to pump the water up from the aquifer and across town. Someday, people will find it hard to believe that their ancestors once irrigated lawns with drinking water, but the practice is still common today.

While neatly-swept yards may no longer confer high social status in the American South, they continue to be an important indicator of functioning families elsewhere in the world. Zimbabwean friends tell me that, in addition to lowering the risk of fires igniting their thatched roofs, the clean sweep around their houses also reduces tsetse fly densities and allows easy monitoring of the comings and goings of visitors – human and otherwise, invited and not. Similarly, given all the grain pounding, wood chopping, and pig rooting that goes on around

Iban longhouses in Borneo, it's no wonder that yard sweeping is still *de rigueur* in that part of the world. Meanwhile, swept yards are scattered around the South; there's a really nice one out at the Dudley Farm State Historic Site on Newberry Road.

The Professional Lawn Care Association of America has proclaimed April as "National Lawn Care Month." Out of concern for our air and aquifer, and given that May is typically a month of droughts and wildfires, perhaps it should be proclaimed the "National Swept Yard Month." Corporate sponsors for this campaign are currently being sought. In any event, given the financial and environmental costs of maintaining lawns, we need to adjust our aesthetics and free ourselves from the clutches of the lawn-care industry.

8. Forests Have Their Place, but These Are Savanna Lands

During the first few years after I moved to Gainesville, I was principally attracted to closed-canopy forests. I considered these forests to be more natural than the pinelands in which they nestled, and I was impressed that they could support up to two-dozen species of canopy trees, including such wonders as cabbage palm, magnolia, and basket oak. Although their diversity is lower than what I had grown accustomed to in Panama and Malaysia, it is much higher than the forests of my youth in New Jersey, Wisconsin, and upstate New York. To get to the Florida forests with their stands of broadleaved trees, I traversed miles of nondescript pinelands. Only later did I realize that many of these so-called "hammocks" of hardwoods are artifacts of mismanagement – testimonies to misunderstanding the ecological roles of fire – and species-poor in comparison with the pine savannas they replaced.

But I was not alone in my mistaken belief that hardwood forests represented the predominant "natural" vegetation of Florida. I put "natural" in quotes because the answer to the question of "What is the natural vegetation of Florida?" is "It depends." As in other parts of our changing world, the vegetation that develops depends principally on where you are geographically, hydrologically, and historically. In contrast to most other places in the world, however, Floridian vegetation is distinct insofar as, almost more than those other factors, it is shaped by fire history. In other words, fire can trump location, flooding regime, and time period in determining whether an area supports hardwood hammock, sandhill, scrub, flatwood, or swamp.

The terminology used to describe vegetation in Florida was mostly new to me, and it deserves some explanation. A hardwood forest in a sea of pines is – as mentioned above – called a "hammock." No one is quite sure about the derivation of this word; perhaps it has Seminole roots or perhaps it comes from "hummock" as in "small hill." Then there are sandhills, which can be quite flat, but are always dominated by scattered longleaf pines in the overstory and a wide diversity of grasses and broad-leaved forbs in the understory. Flatwoods, which are always flat and frequently squishy underfoot, are also pine dominated, but whether the pines grow in open savannas or in more closed-canopy woodlands will vary with the fire regime. The dominant pine species in flatwoods varies with the depth and duration of flooding, with longleaf dominating at the drier ends, pond pine in the wet spots, and slash pine in between.

Where the flooding is deeper and lasts longer, but not so long as to preclude tree establishment, any of a variety of swamp forests develop. Small and isolated depressions in flatwoods develop into bayheads if protected from fire and cypress domes if not. Bayheads are forested wetlands dominated by a variety of evolutionarily unrelated trees, all with evergreen leaves and all known as bays; bull bay, swamp bay, and loblolly bay are the most common. Cypress domes, in contrast, have their tree diversity cropped by occasional fires that leave only thick-barked pond cypress and black gum trees mostly unscathed. Cypress swamps were formerly common in places where there is sometimes deep flooding by flowing water, but, between logging and fire suppression, hardwoods now dominate.

When the flooding is really deep and lasts for months, then prairie vegetation develops. These are not the prairies made famous by either Willa Cather or a family of very nice people living in a little house; rather, these are what folks up north might rightly call a marsh. Prairies gradually become swamp forests unless they are really wet or a fire comes through and re-sets the successional clock. In this parade through our ecosystems, I should have started with one that is found only in Florida, and only on our highest and driest sites. Scrub develops on the severely drained and nutrient-poor paleodunes scattered down the central highlands of the peninsula and on dunes along the coasts.

For thousands of years that only ended with the widespread availability of air conditioners, most of the Southeastern Coastal Plain (from East Texas up to Virginia) was covered with open-canopied, pine-dominated vegetation. Dan Ward, a now-emeritus plant taxonomist at UF, estimated that, prior to development, 72 percent of Florida's uplands supported pine savanna. He came up with this estimate, by the way, by painstakingly cutting out all the different colored patches in the jigsaw puzzle of vegetation types portrayed on a map that had been published by the late John Henry Davis. He then weighed the pieces of map on an analytical balance, determined a weight-area relationship, and did the math. Somehow, I'm pleased that, even after several decades of increasingly technological sophistication in ecology, an estimate made with such a method stands undisputed.

If I were really dedicated to conservation, I would stay in Florida rather than traipsing off to try to save tropical rainforests. After all, less than three percent of Florida's pine savanna remains intact whereas more than 50 percent of tropical rainforests still stand.

Like tropical rainforests, our native pine savannas are exceedingly rich in species, but the diversity is down among the herbaceous plants in the understory, not up in the canopy. Well maintained – by which I mean frequently burned – pine savannas can support over one hundred plant species in an area not much bigger than your neighbor's front lawn! One reason why so much pine savanna has been lost, and why so much of what remains is in jeopardy, is that so few people recognize its diversity and other environmental values.

I always like to have a work-related justifications and excuses for visiting art museums and spending money on books on landscape painting. Earlier in my career, I wrote a book chapter on the topic of jungles in art, literature, and film, but I've yet to write anything substantial about Florida landscape paintings. When I do, it will be very critical, at least about the subjects the artists chose to depict. I recognize that artists do not choose landscapes at random, and that they also have to paint what sells. Nevertheless, based on a sample of nearly three hundred paintings of Florida landscapes, I found only seven of pine savannas. Not only are landscape painters not sampling landscapes at random, they are showing a highly significant statistical bias against pine savannas. When confronted, artists have explained that what sells is the exotic, hence the proliferation of cabbage palms, white sandy beaches, and cypress swamps. Between the landscape painters, South Florida novelists, and Magic Kingdom promoters, it's no wonder that so few people recognize what is natural in Florida.

9. Junk Hammocks

Since joining the University of Florida faculty, I've taught a graduate-level course entitled Ecosystems of Florida each spring semester. By looking over my old lecture notes, I can see the development of my knowledge about and my attachment to this region. Given the emphasis that I initially put on hammocks – remember that those are the islands of broadleaved trees in seas of pines – I obviously had trouble finding my bearings in our conifer-dominated ecosystems. First of all, I had to learn to differentiate among our half-dozen species of pine, not all of which are so easy for me to tell apart, even now. Then I had to come to grips with the vastly different ecologies of slash, sand, pond, spruce, loblolly, and (champion of them all) longleaf pine. Slowly, the emphasis of the course changed from hammocks, which I realized mostly developed only after fire suppression, to pine savannas. This focus can be justified in many ways, but it may suffice to point out that, over the last century, we lost 97 percent of the pine savannas that had formerly blanketed the South. Alarmed as I realized that the rate of pine savanna destruction has not diminished, the course began to emphasize management and restoration of this now-endangered ecosystem.

Much of northern Florida that hasn't been paved over, suburbanized, or converted into pine fiber farms now supports what I often refer to as "junk hammock." I employ that pejorative qualifier to differentiate anthropogenic stands of hardwoods from the more natural hammocks that occupy nutrient-rich soils in the fire shadows of lakes and rivers. When you see pines in the overstory and hardwoods in the understory, you can be pretty sure that you are seeing the combined effects of suburban development and Smokey Bear. Until the scattered "legacy" pines all succumb to the combined effects of lightning and bark beetles, they provide mute testimony to the glory over which they once dominated. My views about junk hammocks might seem extreme, but I've spent years battling laurel oaks and other encroaching hardwoods where they threaten remnants of pine savanna.

To provide the students in my class with "real life" experiences in restoration ecology, I sometimes assign them "clients" who want advice on land management. The clients are mainly friends and acquaintances, but some representatives of government agencies are also included. Most of the land under consideration is "junk hammock." The assignment is for each student to provide his or her clients with a biological description of the property, including a list of the dominant species. Then, based on their "reading" of the landscape, the students provide the landowner with a description of what the area was like when Billie Bartram wandered through botanizing in 1774.

Reading landscapes involves a close examination of the vegetation and soils, consultation of historical aerial photographs, and consideration of the property in a larger geographical context. Trees with large, low branches, for example, are a good indication that, when those trees were young, their surroundings were more open. Species with individuals in the canopy but no regeneration suggests the same. Given that most of our uplands were frequently burned one hundred years ago, it's not surprising that a majority of my students conclude that most of their properties were formerly pine savannas. The next section of each report is a description of what, with the absence of appropriate management, the property will be like in fifty years. With the onslaught of exotic, invasive species coupled with the consequences of continued fire suppression, the students' predictions generally converge on a biological bleakness that is numbing: it's a future filled with junk hammocks and worse.

The final section of each report is a recommended set of restoration activities. All the students (so far) encourage exotic species control and most recommend conducting controlled burns (or using some surrogate treatment, like brush hogging). Their shared goal is to open the hardwood-choked canopy to promote longleaf pine regeneration and to stimulate flowering of the long-suppressed understory plants.

I didn't expect Joanna to give me trouble with this assignment. She was such a good student, always so quiet and polite. How could I have guessed that she would be the one to object to my "pine savannas everywhere and always" credo? The 45-acre property she was assigned to read and interpret is high, dry, and close to Paynes Prairie. Most likely, it once supported longleaf pine savanna. Why would she object to restoring not only longleaf pines, but also some two hundred or more species of associated flowering plants? Could she be so hard-hearted as to not care about the gopher tortoises that need open canopies, or the multitudes of other species that depend on gopher burrows for refuge? And what about the endangered red cockaded woodpeckers that might move in once the big pines are back? Is she unaware of the glories of fox squirrels? Doesn't she like to hear the wind susurrating through pine boughs? Finally, is she immune to the pleasure of watching low, intensity-controlled burns move through the understory? Granted, the property is bounded by I-75 and is otherwise surrounded by condominium complexes, but those are just challenges for the burn boss. Sure, historical records do show that most of the area was plowed and planted in a succession with cotton, peanuts, and other crops before being managed intensively for pasture. And I'll admit that, in the absence of even a single sprig of wiregrass or a rosette of blazing stars, most of the floristic diversity would have to be re-introduced.

Instead of repeating back to me what I later realized had been elevated to dogma, Joanna argued that the money that would be required to restore pine savanna to her assigned area would be better spent on other conservation initiatives. Then she went on to suggest that the conditions on her site had been so drastically altered that pine savanna was no longer a viable option, especially given the site's location in the landscape. Instead, she recommended controlling the exotics out and managing it as a hammock.

The other students in the class and the property owners who had assembled for the oral presentations of their reports found Joanna's recommendation quite reasonable. To save face, I never dropped the pine savanna mantle, but her suggestion got me thinking. Perhaps "junk hammocks" can be enriched by planting basket oaks, magnolias, and any of the other thirty-or-so native canopy tree species in gaps created by removal of the water oaks and laurel oaks that typically dominate. The understory could be enriched with sparkleberries, styrax, horse sugar, and any of the dozen other beautiful shrubs that occur in our natural hammocks. There are even herbs, like jack-in-the-pulpit, that could be planted to biologically brighten the picture. Maybe there's a future for junk hammocks after all. More fundamentally, perhaps I should sometimes work with the conditions confronted, not against them.

My awakening humility vis-à-vis junk hammocks and savanna restoration took a hit when, first, major drought and then two tropical storms in succession swept through our area. First, the drought killed a lot of hardwoods that couldn't tolerate the deep, dry sands where longleaf pines used to thrive. Then, the storms toppled over large numbers of stressed laurel and water oaks that had encroached into the uplands after fires were suppressed. After all that canopy-tree death, formidable tangles of grapevines developed that were made even more impenetrable when they were interlaced with catbriers and blackberries. In contrast, through the same drought and storms, our longleaf pine savannas flourished.

10. Hey Y'All, Watch This!

A while back, I spent a Sunday afternoon working out in the woods with my neighbor, Padgett. He's a prize-winning Southern novelist who likes controlled burns and some other activities involved with ecosystem restoration. Like the rest of Flamingo Hammock's locations, the land near his house was carved out of a former cattle ranch and is subsequently suffering from cattle withdrawal. Like many other people who basically enjoy the South, Padgett likes live oaks, especially those of the broad-crowned, Spanish moss-draped variety. These are the sorts of trees under which Scarlet fumed about Rhett's wayward ways and from which Tom Dooley hung for his. Unfortunately, such trees are not had for the asking, at least not in the absence of cattle. Through their browsing and general lolling about, cattle shaped these trees and the landscapes in which they figured so prominently. But cattle come with the price of fencing and veterinary bills – generally more effort than we are willing to invest. So there we were, doing cows' work by trying to save some beleaguered open-grown live oaks from the onslaught of other, less reputable trees.

After destocking and with no cattle to browse and trample down the water oaks, laurel oaks, sweetgums, and lots of other fast-growing trees, they do exactly that – they grow fast. Within 15-20 years of cattle removal, laurel oaks and their kin can – at least if there are no fires – pierce up through or otherwise encroach upon the live oaks that don't mix well with other trees, especially close neighbors. Within a few more years, all that's left will be live oak skeletons and shady thickets of lesser hardwoods. So with the heartfelt desire to keep this turn of events from turning any further, we were out doing cows' work.

Padgett was working with his bright orange Husquavarna chainsaw. He likes that saw because it's light and usually starts with a few pulls. I don't like it because it sounds like a sewing machine and because it starts. Being red-blooded and all that, I too have a chainsaw, but mine isn't a Husquavarna, and it often doesn't start. It runs fine on starter fluid, but is usually unhappy with a normal gas mix.

We have a perfectly fine chainsaw doctor in Gainesville, but I admit that, in addition to mechanical ineptitude, my ego is bruised easily, especially by tall, dark women with smoky voices. In contrast to my ignorance about internal combustion, April knows everything worth knowing about chainsaws. She also remembers her customers and the reasons for their previous service appointments. I'll take it in for repairs, but I'll want to recover a bit from my last visit, since I didn't do so well at holding up my side of the discussion about the malfunctioning of my saw's carburetor. And don't think that I'm the only chainsaw-toting male in Alachua County who is intimidated by April. I recall one beery discussion about the topic, during which one friend admitted to having purchased from her a larger chainsaw than he needed after she'd mentioned that it was usually the ladies who liked the model he had intended to buy. To delay the bruising that I had coming, and for a complex set of other reasons that I won't elaborate on here, I was now working with my 30-inch bow saw. On trees less than six inches in diameter, it's not much slower than a chainsaw, plus it's quiet, and always starts.

Even if we were using the same tools, Padgett and I would still address the encroached live oak problem differently. Padgett is methodical. He sizes up the tree to be felled, determines the appropriate felling direction, and, often, sets it down where he intended. He makes his cuts down close to the ground and generally does a neat job. I respect him for his conscientiousness, but there we were, out in the middle of the woods, out of sight of any roads, houses, or even trails. We'll burn in there sometimes, but only if we can get enough fire-carrying fuel down near the ground. The biggest trees we were cutting were only a little larger than those I could handle with my bow saw.

So why bend over to make the cuts? Who cares about high stumps? They may be ugly, but no one will see them out where we were working. Furthermore, I believe that sprouts from high stumps are more prone to heartrot and fire damage than basal sprouts. That I am lazy about bending might have occurred to you, but I assure you that I was not cutting trees at a comfortable height out of any concern for comfort. In any event, my approach to culling laurel and water oaks threatening live oaks was quite different from Padgett's.

Among the criteria I use to select weekend activates, fun figures prominently. If I really wanted to do some good for myself, I would be doing my wife's bidding and fixing the bathroom door, replacing the broken faucet handle, or even finishing the taxes. Instead, I was out in the woods with my silent-but-deadly bow saw, felling live-oak encroachers. I was working a good distance from Padgett; you don't want to mess with a man with a chainsaw, even one that sounds like a sewing machine. And I wouldn't want to work close to him anyway because his methodical, tree-by-tree approach might interfere with my very different approach to the same basic challenge.

While never losing sight of the long-term objective of keeping the world safe for live oaks, my short-term goal was to break my own personal record of the number of trees falling at the same time. To some people, such a goal may seem both dangerous and pointless, but, in my defense, I want you to remember that we were dealing with pretty small trees. As for being "pointless," I'm not sure that I have anything to say to people who might harbor that attitude.

To get bunches of trees to fall simultaneously, I went around sawing most, but not all, of the way through the stems of as many trees as I could hope would come crashing down in unison. Then, after a pause to inspect my handiwork, I felled the "kingpin." Sort of like playing with dominoes, but more thrilling. Most of the time, the results of all this effort are disappointing. Either the "king" turned out to be some sort of lesser duke, or the notches weren't deep enough, or I just plain missed. No big deal – finish cutting them down and find another patch of trees to notch. When 5-10 trees fall in one fell swoop, it certainly gets your blood racing. Hard to imagine that a co-worker only a hundred feet away wouldn't notice all the fun I was having and want to partake. If Padgett had noticed, then it's hard to explain his reaction to what I thought would be an uplifting experience.

I could tell that Padgett was either getting physically tired or that his tolerance for our arboreal destruction, even for the good cause of saving live oaks, was weighing on his spirit. I prepared myself for terminating our activities by preparing for him a nice cluster of grapevine laced trees, nicely notched, and with a splendid kingpin. When he turned off his saw, I asked him if he would mind cutting down just one more tree. Earlier, he had suggested that I trade in my bow saw for a motorized model. So, when I asked for help, I believe he smirked a bit. Perhaps I should have warned him about the events that were about to unfold, but we weren't talking much, and I assumed that he was in on the plan anyway. He wasn't. He did what was expected of him. The plan worked well enough to earn him eight notches in his belt. To my surprise, Padgett's reaction was complicated, non-verbal, but not particularly positive

We packed up our tools and went our separate ways. I hope that we will work together again soon. There are plenty more live oaks to save. We had some fun, and were in no more real danger than is generally expected when messing with saws and trees. I have to admit though, that I thought of Padgett's reaction to my surprise when I heard that old redneck joke on the radio:

> "What was the last thing the redneck said before he went to the hospital? 'Hey, y'all, watch this!'"

11. Killer Trees

A while back, Gainesville's Tree Appeals Board, which I then chaired, met to decide the fate of a large laurel oak that hung over a city street in the old part of town. I should have been apprehensive about the meeting, having been badly scorched several times over the years as a result of our decisions, but I anticipated the session with excitement instead.

Many cities have citizen advisory boards for issues related to urban forests, but Gainesville may be unique in that it has a panel of experts who are called in to mediate when the City Arborist disagrees with a citizen over whether a tree should go or stay. Additionally, I appreciated having opportunities to share opinions about trees with real experts, like the late Noel Lake, the planter of most of the now-large trees on UF's campus; Bob Simons, perhaps the best dendrologist in the region; and Joe Durando, an organic farmer, a horticulturalist, and the frequent president of our local chapter of the Florida Native Plant Society. Plus, I always enjoyed professional dealings with the City Arborist at that time, Meg Niederhoffer, who was not only my friend of nearly thirty years, but was also Hutch's wife and, of course, my next-door neighbor.

The tree whose future we were to determine at this scheduled meeting was a huge laurel oak that had been planted about eighty years ago along Main Street in downtown Gainesville. The Arborist hoped to save this last remnant of a once-glorious avenue of laurel oaks, but Florida's Department of Transportation (FDOT) had another fate in mind. Mediating this particular disagreement was unlikely to be as harrowing as when we were caught between the Arborist and the residents of a well-organized neighborhood association that I will not identify, other than to say that their name has something to do with a place birds swim. Certain outspoken members of that association had objected to the removal of several of their magnificent but decaying and seemingly dangerous laurel oaks. That particular controversy died down only after one of the trees that the Tree Appeals Board mistakenly spared fell and squashed a new Saab. Fortunately, Board members weren't libel, and everyone was relieved that a school bus wasn't involved. We were also humbled by our collective failure to accurately predict that tree's fate.

For the first five decades or so of their lives, laurel oaks make great street trees. Fairly tolerant of soil compaction, capable of rapid growth, and evergreen, they are greatly appreciated for the deep shade they cast when they are fully crowned. Unfortunately, laurel oaks remember every affront, from roots cut during sidewalk construction to bark peeled by careless weed-whackers. The affronts are remembered in the form of heartrots caused by fungi that enter through wounds. By the time they reach venerable sizes, most laurel oaks have become hollow. Laurel oaks kill and maim more people in Florida each year than any other tree species, partially because they were so frequently planted on roadsides, but also because of their penchant for dropping huge branches or falling in their entirety.

Many people can't tell laurel oaks from live oaks, but the differences – though often subtle – are consistent. The leaves of the two species are about the same size and shape, but, where live oak leaves are a dull whitish on the underside due to abundant tiny hairs, the underside of laurel oak leaves are shiny. Although laurel oak crowns spread a bit when open grown, they never reach the prodigious proportions of live oaks. The other clues are that laurel oaks are seldom heavily festooned with Spanish moss and their big branches support few resurrection ferns. More to the point, at fifty years old, most laurel oaks are starting to deteriorate whereas, at that age, live oaks are still adolescents.

All over North America, oaks are common colonists of prairies, savannas, and other open-canopied ecosystems, at least after fires are suppressed and livestock is removed. After fire suppression in our area, water oaks invade flatwoods and turkey oaks clog dry sandhills, but the most abundant, most aggressive, and most rapid post-fire suppression invaders are laurel oaks. At home, my battle against laurel oaks is never ending. I cut them down, girdle them, and burn them whenever I can. It's not that I don't respect laurel oak as a species. They have their place in wetland forests, but not where they threaten the persistence of hundreds of flowering herb species in our once-extensive savannas.

Laurel oak as a species is also not endangered by my attacks. The cut stumps sprout prolifically. They also have a great capacity to jump the girdles I cut around their stems with a chainsaw, machete, or ax. Let me explain that I often girdle large laurel oaks to avoid the trouble and damage of felling them, as well as to create woodpecker and flying squirrel habitat. By the time a girdled laurel oak finally falls, it usually has lost all its branches and does little damage when it does plummet down. Given their capacity to jump girdles and survive, even after I've sawn several inches into the trunk, the death throes of laurel oaks are often quite prolonged. Even effectively girdled laurel oaks can survive for several years, drawing sustenance from neighbors they are connected to either clonally or by root grafts.

The septuagenarian street tree whose fate was to be determined by the Tree Appeals Board had suffered its share of affronts. Its huge trunk forked about ten feet from the ground, which suggests that it had broken when young or was perhaps pollarded, which was once the fashion. Whatever the cause, as the two now-massive and upright stems increased in girth and grew together, they trapped their bark between them. Embedded bark, as it's called, provides a great disease court and might account for the obvious heartrot from which this tree was suffering. Then again, the decay might have been caused when large lateral roots were cut during installation of the roadside curb. Who knows how else this tree's integrity had been compromised both above and below ground? In any case, the decay pocket in its trunk was sufficiently large enough to support a substantial cabbage palm seedling, perched up in a crotch well above the ground.

Despite its numerous shortcomings, this laurel oak – the last of its cohort in the neighborhood – seemed sufficiently robust that the Tree Appeals Board agreed to a stay-of-execution. Instead of chopping it down and planting a much smaller tree in its place, we recommended monitoring its status as FDOT carried out its roadwork. When it starts to deteriorate quickly, it should be removed. In the meantime, residents of Gainesville can continue to enjoy the shade of this old timer. I should admit, though, that if I had a new Saab, I'd park it elsewhere.

12. Firewood, Burnpiles, and Shitake Logs

Winter is well over, but our pile of firewood is still substantial. That's a pity, since the leftover laurel oak logs will be punky by next winter. Perhaps I can use some of it for barbecuing. Fortunately, the leftover live oak will still be rock hard for next winter. We burn mostly laurel oak wood, which we have in abundance due to our pine savanna restoration efforts.

As I mentioned, my neighbor Hutch uses my neatly stacked and prepared-in-advance woodpile as further evidence of my inability to shake off the bonds of my Yankee upbringing. He finds it amusing that I take pride in putting up a substantial pile of wood, even if we don't end up burning it all. While Hutch is the only one who has ever put his consternation about my cultural failings into words, it turns out that the rest of the neighbors share his concern. Although I always offer them firewood, there are seldom any takers other than on really cold days. That said, it took many years for me to realize that, in the interest of helping me to seem Southern, they have taken upon themselves to reduce the height of my firewood pile when the cold winds blow and their supplies prove wanting. I suppose I should just accept this as another example of the graciousness of Southerners.

Winter isn't long in Gainesville, but it's still often nice to reduce the chill with a cozy fire. Our wood stove heats much of the house fairly well, but, more importantly, it gives us one area that is toasty warm. If we lived in a dense neighborhood, smoke might be an issue, but we don't, so it's not. Wood burning is also virtuous insofar as the combustion does not add to the accumulation of greenhouse gases that causes global warming. The carbon dioxide released from our burning logs was destined for the atmosphere anyway, just through the slightly slower processes that involve bacteria, fungi, worms, termites, wood-boring beetles, and other decomposers.

We can't use all the wood I cut for firewood while making the world safe for longleaf pines and fox squirrels, so we end up making burn piles. I used to have dedicated spots for burning where I would drag branches and roll logs, but I now seldom use the same burn-pile site twice. Instead, I make lots of smaller piles that, after burning, are great places to plant wiregrass, blazing stars, and other savanna wildflowers. While landscape fires don't heat soil more than half an inch below the surface, burn piles cook it deeply, killing off most of the roots, rhizomes, and buried seeds in the process. The reintroduced savanna plants then enjoy several years of a competition-free growth. The only species that spontaneously recruit to the burn-pile sites include a moss, pokeberry, and a few legumes with seeds that apparently germinate after being heated. Minerals in the wood ash would favor nutrient-loving weeds instead of our native savanna plants, which are adapted to starvation diets. To lessen weed infestations, I scoop off a wheelbarrow full of ash from each burn pile to use as a top dressing in our vegetable garden.

When burn piles are discovered as an art form, I will be a Rembrandt. Perhaps mine are not as large as those Richard assembles for the celebrations after the annual Florida Environmental Law Conference, but they compensate for size with style. My technique is sophisticated and worthy of detailed consideration.

I start my burn pines with a layer of pine boughs, if any are to be had. I save any chunks of fatwood I find for insertion into the already constructed pile. Most of what I have to burn is laurel oak saplings and small trees, too many of which are sprouts from the stumps of laurel oaks that I had previously cut down but didn't herbicide. To facilitate drying and airflow into the fire, I lay the branches and trunks parallel with one another and stack them before leaf drop so as to have plenty of fine fuel.

In landscape fires, one advantage of burning piles of brush over trying to ignite dispersed branches and logs is that pile burn permits are easier to obtain from the Florida Division of Forestry (DOF). Sometimes, it's clear why acreage permits aren't to be had – high winds, extremely low humidity, and the like are dangerous – but, many times, although the conditions seem perfect, all you can get is a permit to burn piles, and, even then, you need heavy equipment on site. I've never been exactly sure what they mean by "heavy equipment," but I've had the sense to never ask. What I do know is that, after a few hours of working, my fire rake gets to feeling mighty heavy. I must also admit that some of my piles are not at all tall, though they cover some fairly substantial real estate. Unfortunately, even after stretching the rules a bit, I am still behind on my controlled burn schedule.

If you have a lot of laurel or water oak wood to get rid of, you might consider using it as a medium for growing shitake, lion's mane, or oyster mushrooms. All you need is some freshly cut trunks a few inches in diameter, a drill, the appropriate mycelia with which to inoculate the drilled holes, and some beeswax to seal over the holes after you've inserted the inoculum. Several firms mail-order the inocula either as fungus-infected wooden dowels or sawdust – both seem to work equally well. After plugging a few dozen holes in each four-foot long log, they need to be stacked and kept wet for a few months. After that, you lean them upright against a taught wire or fence, spray them with water daily, and start harvesting your crop. A single log, four inches in diameter and four feet long, might yield as much as five pounds of mushrooms over two or three years of production. At the bargain price of just $10 per pound for fresh mushrooms, you do the math – your oak-encroached sandhill might be a gold mine. Or so it might appear.

I successfully grew shitakes for several years, though perhaps I should qualify what I mean by success first. Our harvests were more than sufficient for home consumption with plenty leftover to gift. My hours peddling 'shrooms at the farmers' market also paid off to the tune of more than $100. That said, I never actually made a financial profit, even when we figured in my wages at $1.25 per hour, which was the minimum wage back when I earned it. As usual, my marketing skills left a bit to be desired. If I ever venture into that business again, I'll make sure to inoculate more logs so as to enjoy some economies of scale.

My best deal with mushroom cultivation was when a local grower asked if she could harvest laurel oak from our property. I agreed, but with apparent reluctance and only after she promised to give us as many shitakes as we could eat. Such deals are rare and not to be missed. If you need laurel oak wood for any purpose, let's talk. Perhaps we too can reach a mutually satisfactory agreement.

13. Controlled Burns

Like most members of my species, I am fascinated by fire – and frightened and attracted at the same time. While I recognize the tremendous dangers posed by wildfires, the opportunity to conduct controlled burns helped me to start feeling at home here in the sandy lands of the South.

Every semester, my students at UF all get to participate in a controlled burn. I must admit that, when I encounter former students, it is often only the fire that they remember from a semester-long course on plant ecology or botany. In contrast, when I was an undergraduate at the University of Wisconsin in the early 70s, burning was extra-curricular and a lot less controlled. At UF, we burn in class not because it's fun, but because it helps young Floridians learn about fire ecology and management. Until fairly recently, fires swept across the state every couple of years. Fire frequencies plummeted after the US Forest Service sent us Smokey Bear on the heels of the Dixie Crusaders and their Southern Forestry Education Project. The US Forest Service now recognizes the seamy-side of Smokey, but fire suppression remains a major threat to the remnants of natural ecosystems in Florida.

Many of the fires that shaped the ecosystems of Florida over evolutionary time were ignited by lightning in May and June, when thunderstorms are frequent and fuels are dry. The frequency of fires increased about 12,000 years ago, when the first people started to pass through, hunting mastodons and giant ground sloths. But it wasn't until rising sea levels backed up our plumbing 5,000 years ago and permanent lakes started to form that people became a more permanent presence in our neighborhood, and fires became even more frequent. People were also responsible for extending the fire season to include nearly all the months of the year.

Portions of the landscapes of North-Central Florida started to change even more dramatically somewhat about 1,000 years ago, when Timucuan Indians started farming and used fire to clear their fields and to control weeds. At that point in time, the Mexicans and Mississippians – the Timucuans's trading partners – had been farming for millennia, while the local folk had been contented with hunting, gathering, and fishing. With only stone tools and fire to clear fields, and the pest pressures and soil infertility that characterize our region, their reluctance to adopt a farming lifestyle is understandable.

Unfortunately, the Timucuans didn't survive the onslaught of crowd diseases introduced from Europe, not to mention what the effects of warfare and enslavement had on their populations. By the late 1700s, only a handful of them remained, and they opted to leave Florida for Cuba with their Spanish friends rather than face the coming onslaught of Georgians. With the Timucuans gone, Peninsular Florida was open for colonization by Creek Indians arriving from what is now Georgia, Alabama, and Tennessee. Perhaps these colonists, referred to as Seminoles, were more adaptable then the Timucuans with whom they had undoubtedly interacted. In any case, they displayed an impressive affinity for the cattle, horses, and citrus introduced by the Europeans.

The Seminole's approach to cattle ranching involved a lot of burning, as it has for millennia in the Old World where cattle and horses are native. African slaves who escaped to Florida, particularly those from the herding tribes of Ghana, Nigeria, and Benin, improved the animal husbandry practices of their Amerindian hosts. Later yet, the Scotch-Irish who filtered down into Florida – mostly from the Carolinas – brought with them their Celtic traditions of burning and grazing from the badlands of Britain. Here in America, their lifestyle was scorned by the same Anglos and Saxons that had driven them up into the hills in the old country. On our side of the Atlantic, the dirt farmers who scorned the cattle grazers were Philadelphians, Bostonians, and New Yorkers. A century later, this age-old antagonism would flare up big time in major war, but it was only after fence laws were enacted in the 1940s that the landscape was successfully controlled by fire-fearing farmers. A consequence of destocking and curtailment of burning was widespread and rapid hardwood encroachment into pine savannas across the South.

Over the millennia, Southerners have had numerous reasons to burn. They burned to kill ticks and chiggers and to otherwise make walking more pleasant and safer because it was easier to see snakes. Fires were used to drive game and to attract mastodons, or perhaps mammoths, and, later, deer to the fresh forage that sprouts up soon after burns. Fires were ignited in autumn to facilitate hickory nut harvests and in winter, when the winds were more predictable, to protect their palm thatch chickees from wildfires. But they also burned for fun.

Any mention of the recreational nature of fires seems irresponsible, given the deaths, injuries, property destruction, and other suffering caused by wildfires. Although I don't wish to diminish these dangers, we must also recognize the almost-universal fascination humans feel for fire. Even in our increasingly-urbanized society, when charcoal-lighting fluid no longer flares up, we still like to gaze into the eyes of our lovers by the flickering candle light. And who can deny being mesmerized by a wood fire on a cold winter night? Heck, you can even buy DVDs of burning logs to play on your television!

A group of us was sitting out in Richard's party field one chill winter night, drinking beer and gabbing while he burned the bahiagrass pasture around us. He was preparing the field for the wrap-up celebration of the annual environmental law conference put on by UF's Law School. The celebration involves a huge bonfire that is ignited every year in a novel fashion – past ignitions have involved flaming arrows, firecrackers, rivers of flame, and, most impressive, a flaming fatwood log catapulted over the heads of the crowd of lawyers and lawyers-to-be by a home-built trebuchet.

That evening, when Richard was carrying out a much more tame controlled burn, my son, Antonio, was about three years old. Like the rest of us, he was fascinated by fire. Richard was igniting lines of fire through the dry grass with his trusty drip torch, but Antonio must have thought he was being too conservative. In the way of encouragement, he called out several times "more on," which quickly evolved into a group chant of "moron, moron!" Richard complied.

Part of the confusion about whether fires are good or bad derives from the diversity of landscapes being burned and the variety of fire types to which they are subjected. While fires vary a great deal in intensity, it's the cataclysmic, stand-replacing, Bambi-burning fires that, along with the estimates of their associated economic losses, capture the public's attention. More and more frequently, we are suffering really destructive fires here in Florida, which is testimony not only to decades of fire suppression, but also to other forms of land mismanagement and poor land-use planning. In contrast, the frequent fires needed to maintain the pine savannas that once covered most of the uplands of Florida are of such low intensity that they seldom capture the attention of the press. Unfortunately, any lessons learned in 1998 when hundreds of houses burned in Florida were forgotten by 2007, another record fire year.

In areas where fires are frequent, fuel is constantly scarce, so those fires tend to burn with low intensity. This message was reinforced locally during the widespread wildfires of 1998:

where they burned through fire-suppressed, dense pine plantations, the fires were unstoppable. In contrast, the wildfires were easily contained when they moved into fire-maintained portions of the landscape. During that fire year, the City of Waldo was saved from almost-certain destruction by a nicely managed pine savanna on private land at the edge of town.

We don't burn at Flamingo Hammock to drive game, and many of our fires are for ecosystem restoration purposes, but most of the other reasons for burning that were mentioned before do come into play. Some acquaintances believe we burn constantly, but the proliferation of laurel oaks is good evidence that we aren't burning enough. Although it's hard to get a burn permit from the Division of Forestry sometimes, it's often too wet, too windy, or too dry to burn, and it's generally hard to assemble even a skeleton burn crew when it's hot and buggy. We've also made it difficult for ourselves by planting or otherwise encouraging the growth of fire-sensitive plants in areas that would otherwise be easy to burn. It generally takes only a few pulls of a rake to protect a thin-barked tree from serious damage, but a single red buckeye or wild azalea in the middle of an obvious burn unit can be enough to motivate us to take our drip torches elsewhere.

There are too many uncontrollable variables to call controlled burning a science, but there is some science to developing a burn plan. First of all, there are the firebreaks, be they blackened, raked, plowed, or wetted down. Then, there is the decision about the sort of fire to run: a backfire that moves slowly into the wind or a head fire that races along in front of it. Backfires sound like a good way to go, and they do tend to have lower maximum temperatures than head fires, but, because they burn in one place for longer periods of time, they tend to generate more heat. Furthermore, if sloughed bark has accumulated around the bases of big pine trees, and if this "duff" is at all dry, it tends to ignite during back fires, which can then lead to days of smoldering, nasty smoke, and fire girdling of the very trees we are trying to save.

The phrases "controlled burns" and "prescribed fires" are great insofar as they inspire confidence, but every burn boss I know has had at least one fire get away. I've had several, but the only one that did any real damage to property and psyche was when I was a kid back in New Jersey. Our neighborhood in Cranford bordered a fairly large complex of forests and abandoned agricultural fields where we kids played. Those old-fields were maintained by frequent fires that were reportedly started by bad boys from across town. I had always intended to tell my parents and Mr. Allen, our next door neighbor, that I did not deserve praise for dousing the fire that had spread from the fields and ignited his garage. Half a century later, it's too late to admit to my now-deceased parents that I had also ignited that fire, but I suspect that they were well aware of the situation and chose to allow me to be burdened with its guilt.

I worry that children growing up in places like Flamingo Hammock might suffer some cognitive dissonance when they learn about the dangers of fire at school only to come home and watch their parents burning off acres of pinewoods. That dissonance must have started back in the 1930s when The Dixie Crusaders, a US Forest Service-sponsored group of anti-fire evangelists, preached against the evils of piney woods burners. Then there was Smokey Bear, one of the most effective advertising campaigns ever, urging us to stomp out wildfires.

Only a generation or two ago, people in the USA dealt with fire daily. Houses were heated and lighted with fire, which was also used to cook their food. Nowadays, in contrast, our experiences with open flames are few and far between. This lack of familiarization with fire does not happen at Flamingo Hammock, where we instill in our children a healthy respect for it while teaching them about how it behaves. How it can be a useful, if dangerous, tool is a necessary part of that lesson. Though they behave responsibly during our controlled burns, I wonder if they have sense enough to avoid relating their experiences to their friends and teachers, let alone to the uniformed firefighters who visit their classrooms to instruct them about fire safety.

With due regard for the dangers involved, if we are to maintain and restore the ecosystems that dominated the landscapes of Florida for the past millennia, we need more fires, bigger fires, and even sometimes hotter fires. But these fires need to be ignited and controlled by people who have the knowledge and experience needed to assure that the fires serve their restoration purposes. Perhaps some of our children will grow up to be fire ecologists, but I'm sort of hoping that at least a few choose investment banking or some other more lucrative profession.

14. Practicing and Preaching Pine Savanna Restoration

The ecosystems of Flamingo Hammock were in really bad shape when we bought the property. Decades of overgrazing and fire suppression followed by more years of cultivation had left the wooded areas overgrown with grape-festooned laurel oaks. The abandoned pasture, moreover, was not only sprinkled with head-high clumps of dog fennel, it was also interspersed with taller loblolly pines that overtopped the exotic bahiagrass in places where there weren't prickly pear cacti. Our house perches on the edge of one of those former pastures, which was where I started out on the wrong foot of what has turned out to be a long-term effort at ecosystem restoration. I assumed, at first, from the combination of sandy soils and sparse pines, that the soil was acidic as well as nutrient-poor. Busy as I was teaching tropical ecology and building a house, I had little time to attend to my pasture problem or, for that matter, to learn much about the ecosystems of Florida. All this is an excuse for why, when the folks from the local water treatment facility called and offered to deliver twenty cubic yards of slag lime for free, I jumped at the opportunity. I wasn't sure about the "slag" part of the equation, but was pretty sure that lime was just what was needed. The very next day, I dug out my trusty soil pH kit and brought it home, but the lime had already been delivered. To my surprise and then dismay, instead of suffering from acidity, the pasture soil was circumneutral in pH—sweet and not sour, in the vernacular. Absolutely the last thing we needed was lime, but there it was, in five piles of what looked like dirty snow with the consistency of wet potting clay. I had been had. The water treatment folks duped me into serving as their solid waste disposal facility! With no obvious alternative, I set about burying the stuff, pile by pile. The only positive side of this experience was that my excavations revealed that pawpaw roots penetrate down at least eight feet, which was as deep as I was willing to go.

With the lime fiasco behind me, I turned my attention to restoring the pasture to a longleaf pine savanna, an effort that still continues to occupy much of my time many decades later. To be fair, my efforts represent re-creation more than restoration. Although there were some hardy pawpaws and a few other native savanna plant species scattered around, I am basically building a savanna from scratch. In retrospect, it might have been better off had I started by knocking down and plowing up all the existing vegetation and then herbiciding all the weedy colonists for a couple of years. On that clean slate, my plantings of longleaf pine seedlings, wiregrass tublings, and wildflower seeds might have thrived due to the absence of competition from bahiagrass and its nefarious pasture associates. But hindsight is 20/20, and, anyway, who wants to look out over a clearcut or use that much herbicide? Furthermore, I have to consider all the healthful exercise I've had over the years planting and replanting, pulling out dog fennel, grubbing out bahiagrass, thinning out loblolly pines, and digging prickly pears.

If ecosystem restoration is the acid test of ecology, then I have suffered burns over much of my body. On the more positive side, I should also report that it is both humbling and addictive. It is also clear that I am much more effective as a researcher, teacher, and preacher of restoration than I am as a practitioner. Although I claim success with every flowering blazing star and every longleaf pine seedling that emerges into gangly adolescence, I have to admit that the areas where I claim success are rather small and problem prone. When the deer tongue, summer farewell, or goldenrods are in full flower, those patches of hard-won savanna look pretty good, but then there are frequent and humbling surprises that reveal just how little I still know about the ecosystem I am endeavoring to restore. For example, and for reasons that still elude me, one summer, a morning glory vine took over one of the restored patches while I was off doing fieldwork in Indonesia. Another time I did something to stimulate the germination of thousands of dormant seeds of hairy indigo. And then there was that instance when I chanced upon exactly the right fire regime to promote the proliferation of stinging nettles. Such surprises invoke humility, but they also stimulate me and other restorationists to learn more ecology, to try harder, and, otherwise, to tinker in novel ways.

It reflects poorly on my character, but I enjoy bringing students out to see a restored area that I had talked about in class. The twist is that I wait to see which of them is the first to express some statements of truth: "This looks terrible!" or "Is this all?" and so forth. Such realistic reflections only emerge after long uncomfortable pauses, but I hope their discomfort reinforces the point: restoration is really hard.

The patch of restored pine savanna in front of our house served me well midway through my academic career when my department hired a new chair. I invited him and his wife over for dinner, not realizing just how English they are and how worried they were about living in America in general and in the South in particular. Let's just say that, if they went through a book of "you know you're a redneck if..." situations, they would find few that were familiar. In my case, I hit the 50 percent mark. Isn't the proper way to unload a pickup-truck bed full of wood by driving backwards fast and then slamming on the brakes? Overall, my effort to calm their fears didn't quite work out as expected, though it established our relationship on grounds that were oddly favorable for me.

On the afternoon of their visit, I had conducted a fairly large controlled burn but had not gotten to the area along the sand road that leads to our house. Knowing that I had blackened around it already, and in recognition of my so-far neglected duties in the house preparing for our visitors, I sent my daughter Juliana, then nine years old, out with my drip torch to finish the job.

Wouldn't you know it that Juliana was putting down a line of fire just as David, my new boss, and Hazel, his wife, drove up in the Morris they had imported from the Old Country. They rushed into the house blathering quite volubly about a little girl and a wildfire and flames. I tried to calm them by explaining that the young lady with the driptorch was my daughter and that she had plenty of experience with controlled burns. That stopped their speech, but, after that fateful day, they have always treated me with the respect one reserves for unstable individuals.

Coupling the beauty and fabulous diversity of longleaf pine savanna with how little of this once-extensive ecosystem remains intact, any efforts at restoration should be applauded. My restoration efforts continue to be impeded by mistakes, and my battle is an uphill one due to the ferocity of competition from bahiagrass, the density of viable but dormant weed seeds in the soil, and the residual effects of fertilizer applications.

Bahiagrass is an exotic pasture grass that is the dickens to kill. Repeated fires weaken this stoloniferous menace, and there are herbicides that will kill it, but I generally resort to hand weeding it out of the spots where I'm planting pines, wiregrass, or wildflowers. I try to minimize the buried seed problem by disturbing the soil more than is necessary. Residual fertility is a problem because many of our worst weeds – defined as plants growing where they are not wanted – benefit more from soil nutrient abundance than the starvation diets to which native species are adapted. Nitrogen isn't such a problem because it can be depleted by volatilization in fires and removal of hairy indigo, an exotic nitrogen fixer. Residual phosphorus, in contrast, is harder to deplete, but it leaches more readily when the soil's organic matter concentrations are reduced by repeated burning and bahiagrass removal.

Like many naïve restorationists, I was a long-time believer in the *Field of Dreams* myth (i.e., build it and they will come). Forgetting all I knew about seed dispersal and the need for nearby seed sources, I somehow expected that, by thinning the pines down to an appropriate density and then reinstating the historical regime of frequent growing-season fires, my longleaf pine savanna would miraculously reappear in all its glory. Unfortunately, while my savanna-in-progress has the right structure, I have a long way to go before the plant diversity is up to the desired level. In the meantime, I scour the countryside for seeds and celebrate each new successful reintroduction, waiting for the day fox squirrels move back into the neighborhood.

15. Reading the Landscape

When you know enough about the lay of the land, its vegetation, hydrology, and soils, you can start to interpret the hows and whys of its current state enough that you can venture predictions about what it will look like in the future. Learn more about the history of the area and additional insights will emerge about how the flora became so diverse, why some of the trees have such low branches, and how the grapevines got to be such a nuisance. Dig some holes, examining soil textures and colors as you do so, and new ideas will come to mind while old ones are confirmed, revised, or discarded. Talking to elderly neighbors, looking at old air photos, and combing through property records and surveys can provide more clues about what your place was like in the past, how it got to be the way it is today, and how it is likely to change in the future. This process is referred to as "reading the landscape."

"Reading" is not a perfect analogy for what one does when interpreting nature. But perhaps the process of reading and interpreting ancient texts is the most similar, though one does not approach the *Torah* or the *Aeneid* with a shovel in hand, nor is one's nose of much use in the Apocryphal Books of the Bible. Maybe landscapes should be thought of as written in pictorial languages, like Egyptian or Mayan hieroglyphics. For instance, "pine" in this language might connote something about openness, because most pines require a lot of light to get established. "Loblolly pine" might suggest plowed fields or floodplain soils, "longleaf pine" would conjure frequent fires, and "spruce pine" should cause the reader to expect few fires and lots of clay in the subsoil. Put some low branches on any of these pines, and you have cause to believe your landscape was formerly open, even if that's hard to believe given the current density of hardwoods you're actually seeing.

My favorite reading is the landscape near home. Over the years, as I've learned more about the soils and history of this area and began to recognize more plant species, my landscape interpretations change. For example, I was once convinced that, prior to the depredations of European livestock, the naval stores industry, fence laws, fragmentation, and fire suppression, all our land had been covered by longleaf pine savanna. After comfortably maintaining this assumption for more than a decade, I started to wonder about the southern red oaks and mockernut hickories that grew with such abundance in some areas. I discovered that both are indicators of phosphatic clays in the subsoil and that mockernuts were actively managed by Amerindians. Historically, might there have been hardwood-dominated patches amidst the longleafs? I now also have good reason to suspect that, before it was plowed and planted with cotton, sugarcane, or indigo, Flamingo Hammock's low-lying area was intermittently a lake. In other words, the new information caused a radical departure from my initial impression of a sea of longleaf pine savanna.

I've read few books more than a couple of times, but I never tire of reading our landscapes. Somehow, everything I learn about the Southeastern Coastal Plain, global climate change, resource economics, geology, and the thermal properties of bark all feed into my interpretations. Not all the landscape stories I read are happy ones. A forest profile with pines overtopping hardwoods can repeatedly inspire me with melancholy. In contrast, a longleaf pine bolting out of the grass stage is enough to make me smile.

16. Liberating Live Oaks

Until fairly recently, there was little forest in Florida outside of swamps. Instead, most of the uplands were covered with open-canopied savannas dominated by longleaf pines with a few scattered oaks and hickories. The ecosystem-shaping effects of grazing animals – first mammoths and mastodons and, later, cattle and hogs – coupled with fires ignited by lightning, Amerindians and Crackers made sure there were few places where forests could develop. The savannas that had dominated Florida for the few million years that there has been a Florida were species-rich, both in plants and animals. Due to the recent removal of cattle from the landscape and the reductions of fire due to both active suppression and landscape fragmentation, many of these savanna-adapted species are now in trouble. Among the well-known savanna animals in jeopardy are fox squirrels, gopher tortoises, indigo snakes, and red-cockaded woodpeckers. While deforestation threatens the tropics, the natural ecosystems of our sandy lands are being overwhelmed by an unprecedented wave of forestation.

One iconic Southern species threatened by forestation is live oak, *Quercus virginiana*. Naturalist William Bartram described the majestic spreading crowns of live oaks overtopping Seminole villages when he passed through Alachua County in the 1770s. A few years later, the first forest reserve in the USA was established to protect live oaks from live oakers, those roving bands of boat makers who combed the countryside looking for live oak branches the right shapes to serve in the construction of wooden ships.

Live oaks have been a prominent feature of our landscape for a long time. For example, 18,000 years ago, when the last of the Pleistocene continental glaciations was at its maximum, Florida was cooler, drier, and windier than it is today. Analysis of pollen preserved in lake bottoms reveals that our uplands were open oak savannas for which there is no modern analogue in this part of the world.

As a species, our live oak is by no means threatened with extinction, but live oak trees of the venerated open-grown variety are disappearing rapidly, mostly due to encroachment by other hardwoods. As long as fires burned across the landscape every few years and grazing animals otherwise arrested succession, live oaks growing in savannas and pastures were safe. But in as little as two decades after fires cease and cattle are removed, taller-growing, but less fire- and grazing-tolerant hardwoods, especially laurel oak, water oak, and sweetgum, begin to overtop the live oaks. Some of the encroachers actually pierce up through the live oak crowns, though most overtop from the side. Sadly, live oak trees that have lost more than half of their crown to overtopping hardwoods seldom recover, even if liberated.

Savanna-form live oaks present a fundamental quandary for environmentalists and natural area managers. If active management techniques, such as brush-hogging, cattle grazing, and controlled burning, are required to keep open-grown live oaks happy, then in what sense can they be considered "natural?" And if they aren't "natural," then why should the scarce resources available for natural area management be expended on keeping live oaks from being replaced by lesser hardwoods? I'm less caught up in this debate now that I've realized that what is natural and what is cultural can never be fully resolved. In the specific case of savanna live oaks, I'm pretty much convinced that they were an important component in our landscape both during the mastodon years, before humans came onto the scene, and more recently. Whatever the case, live oaks should at least be safe in our parks, where the charter of the

Florida Park Service calls for the land to be managed for all of their natural as well as cultural amenities, so that both bases are covered.

Suburban homeowners with live oaks on their property should pay attention to the "what to do about live oaks" issue. Most people value live oaks for their longevity, strong branches, evergreen foliage, and elegant shapes. Unfortunately, few suburban live oaks aren't being severely encroached upon by other trees. If the encroachers are not removed, or at least severely pruned back, the beloved live oaks won't be long for this world. Even when homeowners realize that their live oaks are in jeopardy, many are reluctant to have the offending trees removed, both because of the expense and because they like trees in general. Unfortunately, if you are especially fond of live oak trees, then you have to do something to keep their crowns free.

Studies of trees and houses in hurricanes suggest that protecting live oaks, even if it means paying a tree surgeon to clear away the encroachers, could be a cost-saving measure. Evidence for this proposition includes the observation that healthy live oaks protect nearby homes during hurricanes by deflecting wind upwards. Then there is the fact that live oak is our most wind-firm tree species, whereas laurel oak, the primary encroacher, is notoriously prone to dropping large branches or falling in its altogether. When hurricane season comes around, you're lucky if you have a live oak next to your house because, in high winds, your tree is likely to catch not only these flying branches, but also other objects such as accelerated garage doors and hurtling patio furniture.

After decades of awareness-building about the evils of deforestation in the tropics, it's about time that other types of ecosystems received more attention. In the USA, Europe, and Australia, closed-canopy forests are expanding at the expense of native woodlands, savannas, and grasslands. To stop this from happening, more people need to recognize that trees are not always desirable and start celebrating their savannas. This aesthetic conversion may be impeded by the symbolic importance of trees in our culture – the tree of knowledge and the tree of life, to cite two familiar metaphors. On the other hand, public appreciation of non-forest vegetation might be readily enhanced if, as some researchers have proposed, we humans evolved in savannas and thus we have a genetic preference for their dappled shade.

17. Poison Ivy, Southern Canebrakes, and the Military-Industrial Complex

"Gonna need an ocean of calamine lotion
You'll be scratching like a hound
The minute you start to mess around
With Poison ivy . . ."
– The Coasters, 1959

As a professor of plant ecology at a major southern university, I make sure that my ventures into ecosystem restoration are based on the best available science. Unfortunately, that basis does not guarantee success. Restoration efforts have a tendency to go haywire, but never so badly as when my attempt at canebrake restoration resulted in the creation of a poison ivy patch of prodigious proportions.

Canebrakes were reportedly widespread across the bottomlands of the American South before they surrendered to the plow. These nearly impenetrable thickets of American bamboo (*Arundinaria gigantea*) could grow from thirty to forty feet tall and fed bison and bears even as they provided critical habitat for Swainson's warbler, which is now exceedingly rare, and Bachman's warbler, which is most likely extinct. Nowadays, canebrakes are few, hence my efforts at their restoration.

Although I conduct most of my formal scientific investigations in the tropics, the "messing around" sort of research that keeps my brain alive and my back strong is conducted in the woods around our house at Flamingo Hammock. I also believe that the neighbors are amused by my scientific endeavors, which they readily describe to visitors (not that I would always recognize what they say about my projects). If they object enough to any of my experimental manipulations that they need to say something, their comments come discretely, indirectly, and with good humor, as expected here in the polite South.

My initial formula for canebrake restoration was based on a few gleaned bits of biology and some field observations. Cane thrives in light and the moist, nutrient-rich soils favored by farmers. One of the only places on our land where we find cane is down by the sinkhole in an occasionally flooded area with black soil.

Compared to historical photographs and descriptions, the cane at Flamingo Hammock is spindly and sparse. This condition, I presumed, was due to the dense shade cast by the abundant water oaks that fire suppression had allowed to invade the otherwise-open loblolly pine stand. As a first step towards canebrake restoration, it therefore seemed logical to lighten the shade. After that, I would remove the accumulated leaf litter and duff with a controlled burn. I was even willing to fertilize to mimic the effects of the cane growing under passenger pigeon roosts.

Cutting and stacking all those water oaks was the work of many mornings, but I admit I was happier dragging those logs to burn piles than I would've been working the exercise machines in a gym. I also look better in work clothes than spandex. I also prefer the company of my dog to that of most people, and respond better to the throaty purr of my chainsaw than I do to Jack LaLanne's encouragement, Jane Fonda's workout tapes, zumba music, or whatever it is that they play in gyms these days. After a few months of blissful log lugging, I was ready for the burn.

With the help of several graduate students and my neighbor, Doug, who is always a willing burn partner, we started out the morning by igniting the piles of logs and branches I had stacked in the canebrake restoration site. When those fires fizzled, Doug and I were desperate – for once, we had a willing crew, full drip torches, heavy equipment on site, and even the official sanction from Florida's Division of Forestry in the form of a controlled burn permit. So Plan B, conjured on the fly, was to burn the much larger lower field where there was more dry fuel – it needed burning anyway.

Several grueling hours and many charred acres later, I dragged myself back to the canebrake restoration site. As I approached through billowing smoke, I realized that the burn pile fires hadn't completely fizzled, and the whole area was now ablaze. I'd ringed it with a raked fire break, and so I was only moderately anxious, but I did call for backup. Fortunately, the fire behaved as intended, burning only up to the fire breaks and no farther.

After that fire I harbored hopes for cane for months, but then had to admit that those hopes were dashed by reality. The combination of canopy opening and burning turned out to be perfect for the proliferation of poison ivy, but not of cane. If restoration is the acid test of ecological understanding, then I should need hospitalization. On the positive side, as someone whose character benefits from frequent doses of humility, the experience reminds me that it is easier to preach (or profess) than to practice restoration. My neighbors and students will enjoy learning of this mishap.

The canebrake restoration area is now completely covered by leaves-of-three-let-it-be poison ivy, and nearly every tree stem is being ascended by its hairy-rooted stems. The putative mother of all stems – two hand-spans around (but "only a dope grabs a hairy rope") – disappears up in the crown of a live oak on the edge of what I'd hoped would be converted into a canebrake. Poison ivy spreads laterally via modified stems that creep along just beneath the soil surface, so I am not sure whether my miraculous patch originated with a single seed or lots of different individuals. The seed crop of the whopper vine alone would be enough to populate the planet.

The poison ivy plantation I created borders a much-frequented path through the Flamingo Hammock's common land. That I have not yet been confronted with my fiasco is more likely a consequence of Southern manners than lack of botanical acumen. Furthermore, several of us have reached the stage of life and hip-socket deterioration when "taking the dogs for a walk" involves an electric golfcart, which reduces the likelihood of direct contact. Unfortunately, infection via our dogs' fur is still a good possibility.

I could eradicate the poison ivy patch with an arsenal of herbicides – death to poison ivy would be only a few hundred squirts away. But applying that much herbicide would almost certainly kill the few straggling sprigs of cane struggling up through the poison ivy tangle, not to mention the potential damage to pig frogs, spring peepers, and southern toads. Instead of chemical warfare, I keep a wide swath through the sward clear, and have learned to appreciate the fine points of my nemesis.

"Leaves of three, let it be" is good to remember, but every other part of a poison ivy plant is dangerous, too: the bark, the roots, the flowers, and the fruits; if you burn poison ivy wood, even the smoke can get you. The trouble-causing compound is an oil called "urishiol," after the Japanese word for lacquer. As little as one nanogram can cause contact dermatitis. That means that 500 people could be caused to itch by the amount of urushiol on the head of a pin. And don't think about storing that pin in a drawer and forgetting about it: urushiol can retain its potency for decades.

Urushiol's mode of action as an allergen provides some clues about which of the many proposed cures and preventative measures make sense. Within minutes of contact with human skin, urushiol oil sinks into the lower epidermis, where it binds to proteins in white blood cells' membranes. The resulting histocompatibility complexes are recognized by patrolling T-cells from the immune system. What has been described up to this point is just the induction phase of the allergic reaction. A subsequent encounter with urushiol is needed to elicit the full-blown response that includes excited T-cells producing cytotoxic enzymes that destroy not only the cells attached to urushiol but also everything in their vicinity, hence the redness, blisters, and dreaded itch. Peak misery of the 85 percent of urishol-sensitive humans may be delayed by up to a week, but, by day two or three, most people are suffering.

There's some good news and some bad news about urushiol. If you wash with soap within 5 minutes of contacting poison ivy, you might remove that dreaded oil before it penetrates. The bad news is that after those few minutes, you are out of luck. Urushiol sinks in fast, and scratching the itch often leads to infections and subsequent scarring, though it won't spread the rash. Unfortunately, given the shelf life of urushiol, oil-contaminated dog fur and clothing can infect even those who avoid the woods.

The realization that my dog can give me poison ivy but doesn't suffer it herself got me thinking: what exactly is urushiol a defense against? Poison ivy leaves are the choice browse for lots of wild beasts, including rabbits and deer, so it seems that most non-human mammals are not sensitive. A published list of poison ivy associates includes more than 100 species of insects, 20 fungi, and 12 mammals. Even if other animals were urishol-sensitive, given the two- to three-day delay between contact and rash appearance, I doubt they would associate the itching with previous encounters with poison ivy plants.

Urushiol production is also not restricted to the genus of poison ivy, poison oak, and poison sumac (*Toxicodendron*). Instead, it is found in many members of the Anacardiaceae family, which includes mangoes (hence "mango mouth"), Brazilian peppers, and some beautiful timber trees in Southeast Asia. The Japanese have long extracted the oil for their painting from the lac tree, *Toxicodendron verncifluum,* a congener with poison ivy. The cosmopolitan distribution of urushiol-producing species suggests some co-evolutionary story involving humans and relatives of poison ivy, but that possibility seems like a stretch. Instead, perhaps urushiol has some effect on an as-yet undiscovered or now-extinct herbivore or pathogen.

Instead of disappearing, poison ivy seems to be getting more common and widespread, even outside of my canebrake restoration area. Incredibly, poison ivy has been introduced to, and is now spreading in, Europe, Asia, and Australia. Might this be the result of a slow form of bio-terrorism? No one has yet come forward with an explanation of how poison ivy made it to the Old World, and I hesitate to point fingers, but it does make you wonder.

While mismanagement certainly contributed to my poison ivy proliferation problem, ultimate responsibility lies with the military-industrial complex. I will justify this accusation in a moment, but, for now, I should admit that just voicing it works wonders on my psyche. I can feel my hair (now sparse) bouncing on my shoulders, my white button-down shirt becoming miraculously tie-dyed, and my fingers forming a peace sign as I recall multiple stanzas of the "Fish" Cheer, and my face crinkles into an illegal smile.

My accusation of the military-industrial complex is based on a fifteen-year, multi-million dollar US Government-funded research project conducted in North Carolina. Every day for several years, researchers dosed patches of loblolly pine forest with nearly double the pre-industrial atmospheric concentration of carbon dioxide. At the current rates of fossil fuel combustion and forest destruction, this extreme treatment anticipates our atmosphere by just a few decades. In a paper published in the *Proceedings of the National Academy of Sciences*, the researchers reported that, though most plant species benefited from carbon dioxide fertilization, poison ivy flourished alarmingly. Not only did it grow more, but its leaves also contained higher concentrations of a more-potent variety of urushiol than what was found in the plants in nearby control areas.

Further evidence for the military-industrial complex's guilt for my poison ivy problem comes from another large-scale and long-term experiment on global climate change. Researchers at Harvard Forest in Massachusetts used heating cables to warm the soil in a forest patch by nine degrees. One of the treatment's effects was to increase rates of leaf litter decomposition, which results in increased mineral nitrogen availability. A major beneficiary of this nitrogen pulse was poison ivy, apparently due to the sorts of fungal symbionts it employs to acquire those nutrients. Global warming, along with carbon dioxide enrichment of the atmosphere, will lead us towards a poison ivy future.

If poison ivy is our future, more attention should be paid to curing the itch or avoiding it altogether. For the itch, there are patent medicines and various steroidal treatments on the market, but the oft-mentioned treatment that interests me most involves eating small bits of poison ivy over a period of several weeks. This approach to preventative medicine, usually claimed as being the inherited wisdom of unspecified Native Americans, makes sense, since minute doses of some allergens can decrease sensitivity. This suggestion's credibility is enhanced by the observation that Japanese lacquer painters occasionally lick their brushes, reportedly to maintain their insensitivity to urushiol. Unfortunately, according to a clinical trial conducted by AMA-certified researchers, insensitivity to poison ivy is not affected by ingestion of its leaves. Instead, they claim that the most likely results are lip rashes and acute gastric distress. In keeping with my hippie roots, I remain a bit dubious about such pronouncements from the Western medical establishment, but I am still not about to start grazing on my poison ivy patch.

When your life is made hellish by poison ivy, you might be tempted to wish for the extinction of the species, but this fate would have grave ecological repercussions. Poison ivy's lipid-rich fruits are critical to migrating song birds that need to build up their fat reserves. On their way to the Caribbean and South America in autumn, many migrants chow heartily on the little grey fruits displayed up in the canopy. Poison ivy provisions its fruits with these energy-rich compounds in exchange for the seed-dispersal service provided by the birds. Grey doesn't seem like a good color for attracting birds, but poison ivy advertises the availability of its fruits with breathtaking autumnal shows of red and yellow leaves referred to as "fruit flags." Look for these bright colors and the attracted cedar waxwings and tanagers next autumn.

Irrespective of our newfound appreciation of poison ivy, what about the canebrakes? I have a few more experiments planned, though I could use some suggestions. Meanwhile, should you want your own poison ivy patch, for reasons unnecessary to explain, I am the go-to guy for advice.

18. Lost Laurels Lead to a Camphor Conundrum

After decades of trying to eliminate exotic invasive species from Flamingo Hammock, I was distressed to encounter a good-sized Asian camphor tree on the slope leading down to the sinkhole behind our house. In addition to controlling exotics in that area, the goal of my restoration efforts is to enhance the native tree diversity that was depleted by logging and years of overgrazing. One of the native tree species I have favored is redbay, a close relative of avocados and the Mediterranean bay leaves (the herb we use in cooking and for which redbay serves as a good substitute). To the dismay of tree lovers, bird watchers, butterfly enthusiasts, and micro-lepidopteran moth aficionados (all nine of them), redbays are dying across their range, which spans the Southeastern Coastal Plain. Actually, the other members of the laurel family, including sassafras and spicebush, are dying too, from "laurel wilt." Laurel wilt is the one-two punch of an introduced Asian ambrosia beetle (*Xyleborus glabratus*) and its fungal accomplice (*Raffaelea lauricola*). Although camphor trees are also in the laurel family, they are immune to laurel wilt, most likely because they coevolved with it in Asia.

On the day of my camphor discovery, I was out in the woods again, machete in one hand, squirt bottle of herbicide in the other, lopping and stump treating the water oaks that were proliferating in response to the death, by laurel wilt, of our redbay trees. By chopping the native but overaggressive water oaks, I liberate native live oaks, hop hornbeams, magnolias, ashes, elms, and anything else that isn't a water oak. That this exotic camphor tree was already too big for my machete meant that I had missed it during several previous campaigns of liberation silviculture. Rather than chance missing it again, I went up to the house to fetch my chainsaw.

On the walk back to the house, I pondered the paradox of doing environmental good with chainsaws and herbicides. My comfort level with these instruments of the Devil may seem strange, but I can mount a ready defense for resorting to such crude tools. Principally, if I just cut down the camphor, it almost certainly will quickly resprout and start fruiting again in a few years. Completely killing the tree would involve repeatedly cutting the sprouts until the stored carbohydrates in the stump and roots were finally exhausted. I actually employ that approach on occasion, but crawling around sprout clusters with a lopper is not nice work. Furthermore, I'm no longer young, and I have lots of other restoration activities to occupy my time, not to mention my day job teaching ecology at the university.

My camphor conundrum – my hesitation about cutting that camphor tree – is a bit out of character. Typically, I rail against exotic invasives and don't hesitate to root out Japanese climbing ferns, Asian rain trees, African air potatoes, and Brazilian peppers. In the case of camphor, in contrast, I find it difficult to ignore the fact that, after laurel wilt wipes out all of our native members of the laurel family, its only representatives will be exotic camphor trees. What isn't clear is the extent to which camphor trees will ecologically replace redbays.

Other than not belonging in Florida due to its recent Asian origin, I must admit that camphor trees have some redeeming qualities. First of all, it's a beautiful tree, with stout, rope swing ready horizontal branches. The deep shade cast by its evergreen leaves is much appreciated in this land of long, hot summers. Then, there are the memories its aromatic bark evokes of chest poultices mothers so lovingly apply to counter their chilrens' sniffles.

All that sentimentality is well and good, but will caterpillars of the glorious black-and-yellow laurel swallowtail butterflies (*Papilio palamedes*) make the switch from redbay to camphor leaves? The several lepidopterists I asked were not encouraging. Worse, they were downright distraught about the fate of *Phyllocnistis subpersea*, a distinctly more obscure beast that is also a redbay specialist. This lovely little leaf-mining moth spends its larval life making squiggly burrows between the leaf surfaces of redbay trees only: so no redbay means no moths of that species. It's anybody's guess what will happen to this moth's parasitoids, which include several species of tiny wasps that live inside the slightly less-tiny moth caterpillars. With all these connections in mind, I reread Ray Bradbury's "Sound of Thunder," which makes me worry about what else, other than a few tree species and their invertebrate associates, is at stake.

Despite their several demonstrated and potential redeeming qualities, camphor trees have the undeniable tendency to be invasive. Instead of staying put where they're planted, they employ the seed dispersal services of a wide range of birds and mammals – who relish the oil-rich, avocado-flavored fruits – to help them invade different landscapes, from swamps to sandhills. In what passes for winter here in north Florida, hungry flocks of robins and cedar waxwings are just two of the trees' noisiest fruit-eating and seed-dispersing visitors. Why should all those hungry animals care that camphor is not native? It tastes like redbay to them.

As a science teacher, I try to present balanced views on environmental topics where reasonable people might disagree. While I pay short shrift to anti-evolutionists and climate-change deniers, I do present several perspectives on the debate about exotic species eradication campaigns. Similarly, when regarding herbicide use, I explain how today's pesticides are far less hazardous than those that caused silent springs, twitching robins, and defoliated Vietnamese forests in the 1960s. I am also careful to point out that the research supporting claims of herbicide benevolence is mostly funded by the companies that produce the compounds in question. I also present the downside of restoration, not only its financial costs, but also the redemptive opportunities it thereby provides developers: if nature can be restored, then why worry about a little environmental destruction?

Contemplation of camphor evokes sadness and strong feelings of frustration about the death of redbays and the extirpation of other local laurels. Back in 2004, when I first became aware of laurel wilt, the problem was restricted to the near vicinity of the port at Savannah, Georgia. I witnessed its impacts while on a field trip to nearby Sapelo Island. Upon returning to Florida, I started a campaign to get the US Forest Service, the Georgia Department of Agriculture, or *anyone* to stop its spread.

The overworked and underpaid civil servants I pestered about doing something to stop the spread of laurel wilt thanked me for my concern and assured me that they were monitoring the situation and researching the problem. Given the widespread dislike of governmental interventions of any sort, their reluctance to act is understandable, despite their lip service to the "precautionary principle." And then there's the tendency to condemn failed actions but to tolerate inaction, even when it results in tragedy. To be fair, it isn't clear what governmental authorities could have done; mounting road blocks with pickup truck inspections in rural Georgia sounds both prohibitively expensive and just a mite dangerous.

And, even if containment efforts were serious, given the fact that just a single gravid female of this rice-grain-sized beetle can start an outbreak, it's hard to imagine being able to keep this beast and its hyper-virulent fungus from spreading. What makes it even worse is that even virgin females of this ambrosia beetle can still lay unfertilized eggs that will develop into males they can mate with to produce plagues of fertile female offspring.

Funding for such a containment effort was also not likely to be forthcoming on economic grounds, even though redbay leaves substitute nicely for their European cousins in pasta sauces and pretty bowls can be carved from its heartwood. Finally, it occurred to me to seek allies among avocado growers in South Florida. My logic was that, since redbay and avocado are closely related – both are members of the *Persea* genus – avocado growers would recognize their vested interests and join the laurel wilt containment campaign.

My efforts to enlist the support of avocado growers for laurel wilt disease containment were for naught. I suspect that I was dismissed as just another shrilly environmentalist trying to cash in on a state of fear. Now, just a few years later, those same growers are scrambling to find resistant genotypes to save their multimillion dollar industry.

Here in the South, the battle against laurel wilt has been lost. But what is being done to keep it from spreading beyond our borders? I shudder to imagine a Mexico without guacamole or the mountains of Central America without quetzals, bellbirds, or guans, all of which subsist principally on the fruits of redbay relatives. Laurel wilt was apparently introduced to Georgia through solid wood packing material; it's too easy to imagine it being transported to Veracruz, Limon, Colon, or Barranquilla in wooden crates full of automobile parts or computer components.

Frustrated as I am about laurel wilt, some ecologists contend that I am stuck with an outmoded mindset. They cajole me to accept that Planet Earth is well into the Anthropocene, that change is unavoidable, and that there is no going back or even staying in place. They encourage me to embrace novel ecosystems and to stop trying to restore what climate change and more general globalization have rendered unviable in our no-analog present. Some have gone so far as to point out that laurel loss is even less likely to provoke a public outcry than when chestnuts, American elms, and, more recently, hemlocks succumbed to exotic pathogens across eastern North America. I see some sense in their arguments, but I can still prevent biotic homogenization of my little bit of Florida, at least if my chainsaw starts.

19. Spring Comes to North Florida

Although I object when northerners comment about Florida's lack of seasons, it's hard to pinpoint a starting date for spring. I suggest we use the first flowering of yellow jessamine (*Gelsemium sempervirens*, Loganiaceae) in late January. The traditional commencement of spring on the equinox in late March doesn't make much sense for us in the South; it's at least too late for potato planters and bird migration monitors. The flowering of red maples and slippery elms in early January could be used to mark the season, but since autumn colors often peak then as well, precious little time would be left for winter.

Yellow jessamine is worth considering as a spring indicator, but there are other reasons it attracts our attention as we pack away our flamingo-skin mukluks and otherwise emerge from the depths of winter. Jessamine grows under a wide variety of conditions, and its large yellow flowers glow brilliantly in hedgerows and swamp margins for several weeks every spring. The plant is a woody vine that twines to climb and propagates itself both with wind-dispersed seeds and by root suckering. Its capacity to sprout from widespread roots makes it easy to propagate but challenging to control in suburban landscapes. Then there are the flowers, which are lovely but treacherous.

Jessamine's treachery derives from its pharmacopoeia of toxins, many of which find their way into its nectar. Our native bumblebees sup unscathed on the nectar of jessamine flowers, but woe to the honeybee that drinks too much of the alkaloid-laced brew. Apparently, during the few centuries since their introduction here by Europeans, honeybees have neither learned to avoid jessamine nor developed tolerance to its cocktail of over forty toxic compounds. Unfortunately for honeybee larvae and unaware bears, jessamine honey is also toxic. To me, the very idea of toxic nectar confuses the birds-and-bees story, but not without cause.

The phenomenon of toxic nectar is well known to apiculturalists and pollination biologists, and has stimulated a great deal of speculation and some science. One explanation of this trait is that it's unavoidable: plants that protect their leaves from herbivores and pathogens with alkaloids, glycosides, or phenolics can't prevent inclusion of these poisons in their nectar. Perhaps, but it's more interesting to explore the possible evolutionary advantages a species might derive from poisoning some of its flower visitors. Another possibility is that the toxins that kill exotic bees also kill fungal spores and bacteria that might otherwise spoil batches of perfectly good nectar. But why is it that just some of those visitors – as in the case of bumblebees – coevolved as pollinators so they enjoy the same compounds that cause others to roll over and die?

An intriguing explanation for selectively toxic nectars is that the toxins actually reward legitimate pollinators, which have evolved resistance, because they keep would-be nectar robbers at bay. But even some of those insects that survive drinking what, for other species, are toxic nectars act drunk after visiting just a few flowers. Instead of staying around to visit lots of flowers on the same plant, they fly off in drunken stupors, which actually might increase cross pollination. That likelihood is enhanced because drunken bees groom themselves less scrupulously and so retain more pollen on their bodies. The pollen they fail to groom away is then deposited on the next stigma after they sober up. I wonder if some of the alkaloids in yellow jessamine are addictive – bumblebees with a jessamine habit would certainly be highly faithful flower visitors.

Yellow jessamine is also noteworthy because it produces two distinctly different types of flowers, both of which are "perfect," which is the botanical term for flowers that are bisexual or hermaphroditic. Look down the throat of a jessamine flower and you'll notice that some plants have long pollen-bearing male organs (stamens) dangling out of the fused petals that form the corolla tube, with the female reproductive parts recessed down below – such flowers are referred to as "thrums." The other morph, called "pin" flowers, has the opposite arrangement: here the sticky stigma is exerted on an elongated style while the stamens are recessed in the corolla tube. Only pollen from pins can fertilize the ovules of thrums and vice-versa. Yellow jessamine benefits from this elaboration because it effectively prevents self-pollination.

Most of the honey commercially sold in North America is produced by a bee species that didn't co-evolve with jessamine and so is poisoned by its nectar. Unfortunately, because worker bees don't drink much of the nectar themselves – since they use it to provision the hive instead – honeycombs can still be laced with jessamine's poisons. Fortunately, humans are unlikely to be poisoned by honey because, soon after jessamine starts to flower, saw palmettos and other copious nectar producers start in on their spring thing. The only people in any real danger of jessamine poisoning are those who misread the label on their tinctures of *Gelsemium sempervirens*, a readily available homeopathic treatment for a wide variety of nervous and muscular disorders.

But climate change is messing with flowering times in Florida. Although overall it's getting hotter, it's actually staying cool longer into the spring. Whenever spring occurs, jessamine will continue to usher in a flurry of flowering, which makes the appearance of its yellow blooms an appropriate indicator of the start of spring.

20. Farming in Stone Age Florida

My back aches from a morning spent preparing the garden for potatoes. Usually, we plant on Valentines' Day, but this year we had to await a shipment of heirloom Andean seed potatoes. With a wife from the Colombian Altiplano, I am accustomed to tuber snobbery. I just hope that the purple lumpy or yellow finger potatoes we ordered thrive in our sandy soils. I am planting them in what I believe is the traditional manner, except that I am using a metal-bladed spade, not a crude tool I've fashioned from a left-handed fighting conch shell and an elm branch. I also admit that, when I initially cleared our garden site, I used a steel ax, not the chert-headed axes with which I've recently had extensive elbow-jarring experiences.

Like most kids, I've always been fascinated with flint arrowheads, but my serious endeavors into experimental archaeology started only recently. First of all, while doing research in a logging concession in Amazonian Bolivia, we discovered huge quantities of pottery shards in an area reputed to be a virgin forest before the recent depredations of loggers.

Actually, my daughter Juliana was the most avid amateur archaeologist in our group, which included my family, some graduate students, and *materos* – local guys who knew the forest well and could do amazing work with machetes. Claudia, Juliana's mother, was conducting field work for her Ph.D. on bark ecology while I continued my studies on logging. The graduate students had projects on fire ecology, soil compaction, and seed dispersal. Juliana, who was eleven years old that memorable field season, was in charge of her brother Antonio, who had just turned two. For the next three years, we continued to spend a few months camped out in that forest, but then Claudia finished her Ph.D. and put her foot down about living in a tent for several months each year.

By the time the research crews trundled back to camp, Juliana really needed to escape from her little brother. Our favorite escape was to walk together down a logging road to a stream where we would take our daily bath. We often stopped on that walk to examine bits of pottery that had been unearthed by the road graders; Juliana had an uncanny ability to find the choicest pieces.

Confronted by all that pottery, I admit to first wondering why prehistoric people would walk so far out into virgin forest to break crockery. For the sake of my reputation, I fortunately never professionally voiced that idea, but I was instead stimulated to learn more about archaeology. I brought my newfound interest in traditional agricultural practices back home to Florida.

It turns out that archaeologists don't all agree about how people farmed in the Americas before Europeans introduced metal tools. It's hard to settle this controversy because, in addition to metal tools, Europeans introduced smallpox and other diseases that wiped out an estimated 95 percent of the Amerindian population. The other problem is that written records of agricultural practices are scarce in the historical literature because the first crossers of the Atlantic weren't farmers, weren't interested in farming, and weren't noting down much about how food crops were grown. It was only several centuries later that flint knappers, the Society of Primitive Technology, and other experimental archaeologists drew attention to the topic of Stone Age farming practices.

The basic debate in which experts around the world are embroiled is whether people regularly cut down large trees with stone axes, burned the slash, farmed the area for a few years, and then abandoned it for a few decades during which soil nutrients were replenished and weed populations diminished. The alternative is that they maintained permanent agricultural clearings through mulching and other soil husbandry techniques. The former method, properly known as "long-fallow swidden agriculture," has long been assumed to be the traditional form of farming. Steel ax-wielding peasant farmers still practice this form of farming in the tropics, but Europeans once used similar techniques as well. As any slash-and-burn farmer will tell you, longer fallows make for better crops. Unfortunately, for Stone Age farmers, long fallows also make for big trees, which all but a few diehard Danish experimental archaeologists agree are excessively hard to cut down with stone axes.

I dove into this controversy with Trey Fletcher, a UF graduate student, with a chert ax head made by legendary knapper Claude Van Order using dimensions provided by Barbara Purdy, an emeritus archaeologist from the Florida Museum of Natural History. Trey probably should have been working on his thesis project, but instead he organized twenty volunteers – ten men and ten women – to help with a study on the effort needed to fell small laurel oak trees with stone or steel axes. Several such comparative studies are available in the literature, but we considered the gender effect and added the novel twist of the feller's assistant bending the stems of the trees being felled. We figured that pre-tensioning the fibers would make trees fall faster, and that this benefit would be relatively greater for stone ax wielders.

In regard to the results of our experiment, let me first make clear that the verb "to cut" should not be used in reference to what one does to a tree with a stone ax. With a stone ax, all you can really do is shred, bludgeon, or otherwise annoy a tree until it is so weak or so harassed that it finally falls over. As found in other studies of this sort, felling trees with stone axes took, on average, twenty to thirty times longer than felling similar-sized trees with steel tools. For example, it took our best feller only four strokes to fell a six-year-old, 3-inch-diameter laurel oak with a steel ax, but 229 blows with a stone ax to fell a tree of the same size. Having an assistant bend over the tree being felled helped substantially, particularly when the volunteers used the stone ax.

Our male and female choppers were equally effective with the stone axe. This result may say something about our particular volunteers, but we think it has more to do with the short, peeling strokes that must be used with stone axes. During our training period, we learned that the full-body roundhouse whacks that are so gratifying with steel axes only serve to dislodge stone ax heads or break the shafts that we so painstakingly fashioned with bicycle inner tubes. Instead, wrist action finesse is needed to get the job done.

Based on how much my elbows hurt after felling a five-inch diameter laurel oak with my stone ax, I doubt that Stone Age farmers in Florida let their fields fallow for more than a few years. Instead, I expect that, once they managed to clear a field, they kept it in production both by mulching and by infield burning palm fronds and other combustibles brought in from surrounding areas. I also now understand why farming in Florida started only 1,000 years ago, at least 3,000 years after the commencement of agriculture elsewhere in the Americas. I too would rather fish than farm these sandy soils, especially if crop cultivation involves pulverizing trees with stone tools.

21. Passing of the Tung Blossom Queen

More than half a century has passed since the last Tung Blossom Queen was paraded down University Avenue in a convertible Cadillac, but the flowers that adorned her cavalcade still grace us each spring with their pastel presence. The scattered tung trees are but a remnant of what was once a vigorous tung oil industry in Florida. The history of this industry is an intriguing tale full of price supports, import quotas, world wars, submarines, zinc deficiencies, unscrupulous promoters, late frosts, and destructive hurricanes.

The biological source of tung oil is the seed of *Aleurites fordii*, a small tree native to China and in the Euphorbiaceae. Tung's large, heart-shaped leaves ("tung" means heart in Chinese) have two prominent glands at the apex of the leaf stalk. Almost two inches in diameter, tung blossoms are white to pink with yellow nectar guides lining their petals. The fruits are apple sized and usually contain five seeds inside a leathery rind.

Tung oil has been used for thousands of years in China because, like linseed oil, it dries to form a hard, durable, water-resistant surface. The Chinese used tung oil not only as a finish on wooden products of all sorts, but also to waterproof paper umbrellas or, mixed with hemp, to caulk junks. In the US, tung oil was and continues to be principally used in paints, lacquers, and varnishes, but it has also been used as an electrical insulator and as an ingredient in the manufacture of linoleum, oilcloth, and printing inks, including India ink.

The tung oil industry in the US began in the early 20th Century, when seeds from the Yangtze Valley were planted at the Plant Introduction Garden in Chico, California. From there, seedlings were distributed throughout the Southeast. The Florida Agricultural Experiment Station in Gainesville was a center for tung research, and several prominent local landowners were major promoters of the domestic tung oil industry. Big-time commercial tung oil production started near Gainesville when, in 1930, the Alachua Tung Oil Company filled the first tank car with American tung oil. In 1950, there were about ten million tung trees planted in Florida, Alabama, Mississippi, and Louisiana. At its peak in the late 1950s, total production in the USA was about eighty million pounds of nuts, which yielded about twelve million pounds of oil. Nowadays, domestic tung oil production amounts to little more than a trickle, and young women no longer aspire to being crowned Tung Blossom Queens.

Tung flowers in Spring, and the fruits ripen in October-November. Sustained production required fertilization with zinc, among other elements, and the short-lived trees needed to be replaced every thirty years or so. In commercial operations, the fallen fruits were left on the ground to dry for several weeks before being collected by hand, sacked, and transported to an oil extraction mill, of which there were once six in Florida. A lot of tung nuts were collected by prisoners loaned to farmers – an easy arrangement to make, since jails were administered by the Department of Agriculture. On weekends and school holidays, grove owners also gathered workers in county-owned school buses that were sent on recruitment trips to rural towns populated mostly by African Americans. Friends in Windsor remembered those festive days, picking up tung nuts for Christmas money.

Prospective tung planters in Florida were encouraged by state-sponsored publications such as "Tung Oil: An Essential Defense Industry" (1942) and "The Tung Oil Industry in Florida" (1945). In the last in a long series of agricultural bulletins (1959), Florida farmers were assured that "during the past 50 years since tung was planted in Florida, the industry has prospered and expanded ... the tung orchardist is now a substantial and satisfied citizen."

Contrary to such pronouncements, commercial tung cultivation failed in the USA because of climate, competition from other oils, and high labor costs. Near Gainesville, late spring frosts reduced fruit production to almost nil in three out of five years. Genetically improved trees were late blooming and thus less susceptible to frost damage than those planted in 1930s and 1940s, but the cost of replanting was too much for many plantation owners. Furthermore, when the improved trees became available, tung growers were still staggering from the effects of Hurricanes Betsy (1965) and Camille (1969). Finally, the toxic properties of tung make it a less versatile crop than linseed. When the market value of drying oils is low, linseeds and soybeans can still be used to make high quality feed, whereas tung nuts are simply left to rot.

If tung oil production suffered from so many problems, why was it so zealously promoted? One reason was that, during the depression, southern agriculture was judged as lacking vigor and tung was seen as an opportunity to diversify. Another factor that rendered domestic production attractive was frequent limitations on imported supplies. Until a steady domestic source was secured, manufacturers remained reluctant to incorporate substantial tung oil into their paint formulas.

With the outbreak of World War II, supply lines for Chinese tung oil were severed and the war effort was in jeopardy. Among its many uses, tung oil was needed as an insulator for electrical cables in submarines. Several large landowners stepped up to the plate, including several high-ranking UF administrators and at least one president (after whom a tung-nut-shaped athletic dome is named). Patriotism alone might have sufficed, but skyrocketing prices also contributed to the enthusiasm for planting tung trees. Unfortunately for Floridian farmers, growers in Argentina and Paraguay also responded to the price signals.

Although only scattered tung trees now remain in Florida, powerful growers managed to prolong the life of the ill-fated domestic tung oil industry just as Big Sugar is doing today down in the Everglades. The activities of this aggressive group ran the gamut from tung blossom festivals and price supports, to federal crop insurance and import quotas. The Tung Festival Parade of 1931, for example, included fifty-two floats (one of which carried the Tung Blossom Queen and her court), a squad of motorcycle police, and the University of Florida Band. I was fortunate to have had the opportunity to interview that particular Queen at her home in Brooker. Her rambling Cracker-style mansion, complete with antimacassars on horsehair loveseats, sat in the middle of what had been her Pappy's tung orchard. I suspect that her father didn't make his money on tung oil alone – she did mention in passing a relation to the Kennedys – but the industry and the parade did leave deep impressions. And lest you suspect that her being crowned Tung Blossom Queen had something to do with family connections, her appearance and demeanor even more than sixty years later argued otherwise.

My tung research culminated with a most peculiar interview with a senior council at UF – a colleague and friend of many local luminaries, including university presidents and tung plantation owners. Imagine a tongue-tied lawyer! We met in his posh office in the administration building where, for the first fifteen minutes, he was quite forthcoming about the inner workings of the tung oil industry. I was fascinated to learn how many senior university officials were involved with tung. All was going swimmingly until I ventured to inquire about a rumor I'd heard several times that linked a UF official with sales of tung oil to the Germans before we'd formally entered the war but after such sales were disallowed. Apparently, their supplies from China had been cut off, and they needed the drying oils to insulate electrical cables in their U-boats. Instead of answering my question, he swiftly terminated the interview, and I found myself outside of his office door looking at the surprised face of his secretary. Perhaps he had another meeting scheduled.

The demand for tung oil was great into the1960s; although the harvest of 1958 was the largest ever, it still fell short of demand. If this level of production could have been sustained, we might still be enjoying tung blossom parades. Unfortunately, paint manufacturers became unwilling to vary their mixtures from year to year in response to the windstorms or late frosts that caused fluctuations in domestic tung oil availability.

The downward spiral of the domestic tung oil industry accelerated in 1966, when the import quota was removed. It does seem odd that the tariff on imported tung oil had increased to twenty-four cents per pound when the world market price was only fourteen cents. Then came Hurricane Camille and the end of tung oil production in Florida.

This Spring, when the tung trees are in their flowering glory, look on them as the Edsels of US agriculture and remember that tung had its day.

22. Sparkleberry Carrs and Oak Domes

I spent the morning crawling through sparkleberry carrs, freeing them from encroaching laurel oaks and various vines. If you have trouble picturing just what I was up to, note that a "carr" is a dense cluster of shrubs. The word is derived from the Old Norse "kjarr," for underbrush or shrubs. A shrub, by the way, is a multiple-stemmed woody plant. Near home, we host carrs of winged sumac and of crookedwood, but the sparkleberry carrs are the most deserving of the name. And again take note that I show my good breeding by calling them sparkleberry and not farkleberry, but I won't dispute that farkling is a noteworthy effect of eating too many of this seed-filled fruit.

Sparkleberry fruits look like blueberries, which they are closely related to in every way but taste. Birds, bears, and other animals like them though, and the plant is very attractive. In full flower, sparkleberries look like low clouds. Later, after the bumblebees and other assorted visitors to its flowers have knocked the urn-shaped corollas to the ground, newly fallen snow is brought to mind, even on a warm day in May.

The crooked stems in sparkleberry carrs are closely spaced, and the mostly evergreen foliage casts dense shade, but laurel oaks, sweetgums, and, occasionally, other hardwoods can eventually beat their way through and overtop them. Prolonging the life of our sparkleberry carrs was the reason for my crawling around, loppers in hand.

Without my assistance, I'm not exactly sure how sparkleberry carrs protect themselves from trees and vines. I know that dense stands of shrubs are intense competitors for light, and even more intense competitors for water and nutrients belowground, but it would seem that all carrs should succumb to taller-growing and climbing species eventually. Perhaps the small mammals that enjoy protection from predatory birds by hanging out in shrub carrs provide the service of girdling the encroaching tree's seedlings? Meadow voles are known to do this in shrub carrs on power-line right-of -ways up north, so the idea isn't completely without support. Power-line managers in those northlands may not appreciate the voles, but they still do all they can to favor the shrubs that slow the rates at which trees grow up into their high-voltage wires.

On the pine islands in Ocala National Forest and in other well-developed pine savannas, there are often scattered carrs of sand live oak, myrtle oak, and other hardwoods. These patches are dome-shaped in profile, which is why they are referred to as "oak domes." In the course of conducting her thesis research on the genesis of oak domes, Denise Guerin and I had occasion to excavate several. We were surprised that, even in domes with thigh-thick trunks, all the stems were still connected belowground, having presumably developed from one acorn a century or more ago. It was also surprising to find that each dome had only one taproot – a large one – that penetrated down to the water table and supplied all of the leaves in times of drought. By counting the annual growth rings in the stems of dome oaks of various sizes, Denise surmised that the waist-high domes were just waiting for a fire-free interval of a decade's duration to grow large enough to avoid being top killed by the next fire. And if the fire comes too soon, never mind, since the root system just gets bigger, and the post-fire stem growth is just that much faster.

By Denise's calculations, most of the bigger oak domes on Riverside Island in Ocala National Forest got established during the 1930s, when the Forest Service was still fighting fires with evangelistic fervor. Now that the islands are once again burned regularly, the tree crowns in these oak domes are high enough to avoid scorching, the bark is thick enough to avoid cambial damage, and the shade they cast is deep enough to prevent flammable grass growth. Many fires burn around the outside of well-established domes, leaving the fallen oak litter uncharred. The smaller domes are burnt to the ground, but the root systems still survive unscathed, ready to sprout up for another shot at domeness and reproductive maturity.

Perhaps sparkleberry carrs in hardwood hammocks do the oak dome dance with fire. Even if the carr is top killed, the root system survives and the stems sprout back more rapidly and more densely than the trees that threaten them. We don't have enough carrs on Flamingo Hammock to risk testing this hypothesis, and I really don't mind my mornings of liberation, but it really does seem like an idea worth testing.

23. Virgin Forest on the Smoky Mountains over Africa in Florida Where Clearcutting Is Virtuous

Almost every year, usually in February or March, I host a group of visiting ecologists from as far away as Germany, but they're always from the north. Their expressed purpose for visiting is to learn all they can about the ecosystems of Florida, but it seems more likely that they are just escaping inclement weather. Such visitors are sometimes referred to as "snow birds," but they might more appropriately be called "slush birds" or "intermittently frozen mud birds." Whatever you call them, it's always a challenge to figure out what to show them. Sometimes I take them home and conduct a controlled burn, which is a memorable experience, especially for the rather staid visitors from Europe.

I suspect that the opportunity to conduct a controlled burn leaves a big impression, but the cultural experience of interacting with my neighbors who help with the burns is reportedly also memorable. Although most of my burn buddies have advanced degrees, and all once had high-powered jobs, they now "work" at home while their wives go off each morning to salaried positions. In spite of their varied backgrounds, they can all sound like they grew up with Pogo and once dated Miss Ma'm'selle when they interact with visitors – colorful is what they are, to be sure. I also offer visiting slushbirds a trip to sand pine scrub, but I don't advertise it as such. Instead, I invite them for a day trip during which they will walk over the Smoky Mountains and through virgin forest, while peering down towards Africa. I also promise that they will see a variety of rare and endemic species, and I suggest that they might just become convinced of the environmental virtues of clearcutting. They usually doubt that any of this is possible, but then they've never had the pleasure of spending time with Katie Greenberg in the Big Scrub, also known as Ocala National Forest.

For her Ph.D. research, Katie took on the issue of whether clearcutting mimics the impacts of fires that historically burned sand pine scrubs to the ground every fifteen to sixty years. She spent several years in Ocala National Forest identifying and measuring plants, live-trapping mammals, counting beetle mounds, and pitfall trapping reptiles and amphibians (and, once, a baby spotted skunk). Based on her data, Katie came to the conclusion that, at least for biodiversity, clearcutting is a pretty good mimic for fire in sand pine scrub. To understand how such a shocking conclusion is possible, and to fathom the business about the Smoky Mountains and Africa, it might help to provide some background.

Like most of North Florida, Ocala National Forest is capped with sands eroded down from the Smokies. During the Pleistocene – in its last two million years – the sand was moved around and created the rolling dunes that are obvious at 55 MPH as you cross the forest. Under this sand, and under the mile or so of underlying limestone, is the complex of rocks taken from West Africa when Pangea split up. So, mountains, at least flattened ones, and Africa, at least of the buried variety, are indeed accessible, but what about the virgin forest?

Lots of people drive through the Big Scrub on Routes 40 or 19, but few ever stop to consider the green blur passing by their windows. I hope they watch for bears, because there are lots. Visitors who stop usually do so in one of the half-dozen picturesque longleaf pine savannas that are like islands in a sea of sand pine scrub. Granted, the pine islands are inviting and make for easy walking, but the scrub deserves more careful consideration. Furthermore, longleaf pine savannas are, or at least were, found all across the Southeastern Coastal Plain, whereas sand pine scrub is endemic to Florida and found nowhere else. And while the frequent controlled burns and other management activities have made the pine islands a flowerful delight, they were all at least heavily grazed until the 1930s or later.

In contrast to the heavily used pine islands, the scrub was so dry, nutrient poor, and otherwise inhospitable that it provided nothing for cattle to eat and no reason for anyone to bother clearing it for agriculture. Nowadays, most of the scrub is actively managed for pine fiber, but some patches that never felt an ax, chainsaw, or feller buncher remain. Sand pine trees, some of which have reached the ripe old age of eighty years, dominate these patches. Given that sand pine trees are reproductively mature starting around their fifth year, are ready to harvest at about twenty-five, and are dying from old age when they reach their fifties, an eighty-year-old tree is indeed worthy of veneration. Accordingly, sand pine stands with trees that are more than forty years old satisfy the US Forest Service's official definition of "old growth." Contrast this with forests dominated by Douglas fir or western hemlock, in which some trees must be more than 300 years old to qualify as old growth. Perhaps that is why some people are underimpressed by old growth sand pine scrub.

The crown fires that burn through the scrub at long intervals are quite different from the frequent fires that burn through the understories of longleaf pine savannas every year or three. Fire science, which is replete with highly technical jargon for describing fire behavior, captures this difference. A typical scrub fire, for example, might be classified as a "humdinger" or even a "ripsnorter." The 1935 Big Scrub fire that burned 35,000 acres in four hours definitely qualified as a ripsnorter; it reportedly bounded across the scrub faster than a horse could run.

As Marjorie Kinnan Rawlings described in *The Yearling*, life is rough in the Big Scrub. Most of the plants have tough evergreen leaves, stiff branches, and gnarled trunks, which makes for slow and painful walking. The lack of grass in the understory means that there is little in the way of fire-carrying fine fuels, so scrub is hard to ignite. In fact, most fires that rip across the savannas on the longleaf pine islands go out when they reach the scrub, thus its apt nickname: "asbestos vegetation."

Although scrub is hard to ignite, we still presume (based on the life history of sand pine and associated species) that, over evolutionary time, it burned to the ground in 50-60 year intervals. If a fire occurs before 10 years have elapsed since the last one, the sand pine might not have enough seeds stored to regenerate. If fires are delayed beyond 60 years, sand pine regeneration may again fail due to the scarcity of viable seeds in the cones on the moribund trees. Other than sand pine and sand rosemary, nearly all other scrub species resprout from underground stems or roots after fires.

Although sand pine trees are killed by fire, the species regenerates abundantly from dormant seeds held up in the canopy in closed cones. The scales of these serotinous cones are stuck together with resin, which becomes brittle after being heated by a fire. A few days later, the scales snap apart, releasing the small, winged seeds. Postfire seed rains in sand pine scrub are impressive, with several million seeds falling per acre.

A few months after joining the faculty at UF, I was asked by an environmental group to help them with their campaign to stop clearcutting on public land in Florida. With my recent acquisition of a Ph.D. and a professorship, I was undoubtedly quite full of myself and more than willing to provide expert advice on a range of topics, including those about which I didn't have a clue. As a longtime opponent of clearcutting, I am surprised that I held off joining the campaign until I could talk with someone who actually knew something about the matter as it related to sand pine scrub. I called John Fitzpatrick, a world authority on scrub jays who, at that time, was at Archbold Biological Station.

John was relieved that I called to ask about clearcutting because he viewed that practice as the salvation for scrub jays, at least in the Big Scrub. He informed me that the acorn-eating jays nest equally well in areas with large patches of vegetation that are recovering from either fires or clearcuts. In smaller patches of acorn-producing habitat, the jays are susceptible to hawks swooping down on them from patch-edge perches.

While I continue to advocate alternatives to clearcutting in slash pine forests and elsewhere, I am in favor of clearcutting scrub, except where high-intensity controlled burns are possible. According to Katie Greenberg's research, even mechanical soil damage by harvesting equipment favors the regeneration of several rare and endemic plant species in the scrub, such as Lewton's milkwort and bonamia. Overall, managing sand pine for fiber seems like an ecologically good fit, except where pine densities are unnaturally high due to severe site-preparation treatments and overseeding.

24. Wiregrass, Lightning, and Cows

My order of 1,000 wiregrass tublings is ready to be picked up from the Division of Forestry Nursery out in Chiefland. At only $125 per 1,000 for the two-year-old plants, it seemed like a bargain, but, now that I have to plant each of them, I sort of wonder. Fortunately, unlike the notoriously fussy longleaf pine seedlings, wiregrass plugs are forgiving, so family members and others can be enlisted into the planting force.

All it takes to plant wiregrass tublings is a 1-inch pipe and some time: push the pipe into the ground, pull out a plug of soil, and insert the tubling. A few heel stomps around the plant and you can be on your way to the next planting spot. If it doesn't rain and your planting area is small, watering the transplants is a good idea, but it isn't usually necessary. For full coverage, you'll need about four plants per square yard, but we tend to plant more sparsely in openings created by pulling up bahiagrass, dogfennel, or hairy indigo plants. Another favored planting spot is the cleared area where the torching of piles of brush has killed the competition.

After two to three years in full sun and otherwise favorable conditions, each wiregrass plant will occupy about as many square feet of ground; if closely spaced, they'll look like so many leaf-spewing fountains. Unlike the aggressive Asian invader cogongrass, or even the familiar South American bahiagrass, wiregrass doesn't spread extensively with stolons or rhizomes. Instead, when a wiregrass plant gets established, it holds its spot, enriching the underlying soil with its root exudates and protecting its buds from fire damage and all but the hungriest of grazing animals. Impressively, wiregrass also survives for many decades after fires have been suppressed and oaks and other hardwoods have invaded.

Wiregrass is adapted to our sandy, nutrient-poor, severely drained soils. While the leaves of its evolutionary predecessors were flat, wiregrass has evolved so that their leaves are now rounded and hollow, the result of first rolling and then fusion of the two leaf margins. At least partially due to this peculiar morphology, wiregrass leaves are tough. Cows eat wiregrass only as a last resort and, even then, only when freshly resprouted after a burn.

The flammability of wiregrass is famous – its clumps should be featured in Boy Scout manuals on Nature's perfectly set fire. When wiregrass burns, it does so quickly, toasting nearby fire-sensitive plants without suffering fatal fire damage itself. An indication that most fires in Florida historically occurred during the summer is that wiregrass only produces viable seeds after an early summer fire; winter burns may knock back its fire-susceptible competitors, but it won't stimulate wiregrass to produce good seeds. Given that lightning is rare during our winters, this connection to summer fire is to be expected, even if over the past 10,000 years of human occupancy, winter fires were also common.

Wiregrass doesn't form a turf suitable for croquet, but it makes a beautiful, zero-maintenance, shin-high groundcover. With growing concerns about fertilizers and lawn chemicals leaching down into our water supply and deteriorating our natural springs, wiregrass seems like a good fit for Florida. Furthermore, there's lots of room between the wiregrass clumps for blazing stars, elephant's foot, and lots of other wildflowers. And while you're at it, enrich the flora with other native clump-forming species, such as lopsided Indian grass, one of my favorites.

25. The Cow Debate

I want to reintroduce cattle to Flamingo Hammock. I believe that we *need* cows because, despite heroic efforts at keeping laurel oaks and other invasive hardwoods at bay with controlled burns, sharp tools, and herbicides, we continue to lose our pine savannas to forests. I realize that cows mostly graze on fresh grass, but, when pastures are poor, cattle will also browse on the vines and tree seedlings that are the bane of my existence. Cows stomping around would also help to keep the understory free for the light-demanding pine savanna grasses and herbaceous plants we want to encourage. I'm not talking about cows of the super-fancy, veterinarian-expecting, hormone-enhanced, fussy-eater variety. The cows I have in mind are the rangy Cracker cows that Lee Gramling so clearly describes in his Cracker Westerns (from which, along with many hours watching cowboy movies, I learned much of what I know about animal husbandry).

Although the cow plan has support from several of the male members of our land trust, Claudia, my wife, objects with some vehemence. She claims that because she grew up on a cattle ranch in Colombia, she can justifiably claim superior knowledge about cows. I'll grant her some advantage, but I need to point out that her experience is not with cracker cows. She just doesn't understand that the breed that made Florida famous for its beef during the Civil War does not need to be molly-coddled. Perhaps more labor will be involved than I expect, and I'll admit that I have a bit more to learn about bovids, but she should also acknowledge the power of my arguments.

Spaniards brought cattle and hogs to Florida in the early 1500s. About twelve minutes after they landed, the hogs went wild and morphed into the "pineywoods rooters" that continue to plague our nature parks to this day. The Spaniard's cows became cracker cows a short while later, to the initial alarm and later delight of the locals. Up until the early 1950s, when Florida's fence laws were enacted and the open range was closed, cows were everywhere. Many of the Southern landscapes that we have come to cherish, especially those that include open-grown live oaks draped with Spanish moss, owe their existence to cattle and the efforts that people went through to keep their cows happy. Now that pasture-raised cattle are rare in this part of the world, hardwoods are encroaching, live oaks are disappearing, and crime rates are increasing.

Cows may be relatively new to Florida, but cowlike beasts coevolved with our native biota and shaped our ecosystems. Mastodons, glyptodonts, and giant ground sloths might object to being likened to cows, but they did have cowlike influences and abounded here until about 10,000 years ago. For millions of years, longer than much of Florida was exposed above the sea, these and other large grazing and browsing beasts roamed Southern landscapes, ate the same plant species that my dream cows would eat, dispersed the same seeds, and otherwise determined which species flourished where. In the absence of any evidence to the contrary, I would go so far as to argue that the bark-eating proclivities of shovel-tusked gomphotheres, and the like, once kept laurel oaks at bay. Due to the current unavailability of gomphotheres, or any other proboscidians for that matter, I will happily settle for cracker cows. Sheep might also do the trick, but our area is getting increasingly exurbanized, if not suburbanized, and too many neighbors keep large dogs with a taste for mutton.

Admittedly, except for deer and a few bison, Florida had been large-animal free between the megafaunal extinction of 10,000 years ago and the arrival of European beasts 500 years ago. This brief interval is but a heartbeat in evolutionary time. By reintroducing cows to Flamingo Hammock, we will re-establish evolutionary relationships that date back to a time before humans were even a flicker in some simian's imagination.

While I mourn the conversion of tropical rainforests into cattle pastures, I have to admit that cows are remarkable beasts. With the assistance of bacterial symbionts in their rumen, they convert fiber (which, for us, is indigestible) into milk and meat. They do this rain or shine, winter and summer, and then walk themselves to market. No wonder cows are the dream animals for subsistence farmers in many parts of the developing world.

Partially in response to Claudia's continued and vocal opposition to my bovine proposal, I called a fellow who owns cattle that he grazes on other people's land. I wondered whether we could at least break even by renting out our land, getting a more favorable agricultural assessment from the taxman, and dealing with our hardwood encroachment problem without having to buy any cows and therefore without exposing ourselves to the veterinary bills about which Claudia is wont to rant. Our conversation went well at first; he was intrigued by my offer to rent him grazing land. Unfortunately, his enthusiasm waned when he saw our so-called pastures and the sorry state of our fences. Apparently, his cows aren't partial to prickly pears and dog fennel, let alone laurel oaks.

While we remain cowless, the debate still occasionally flares up around our house, mostly as a springboard into discussions of what we envision as the "natural" landscapes of Florida. I happen to like live oaks and have come to prefer pine savannas to closed forests, but I admit that both live oaks and open-grown pines may very well need humans – if not both humans and cows – for their persistence. Perhaps the heavy dominance of wiregrass to our few remnant savannas is also an artifact of too many cows, but many of the characteristics of our native species reflect an evolutionary history of grazing and browsing by animals larger than gopher tortoises and white-tailed deer. While cattle might be a bad idea for Flamingo Hammock, I hope that cows continue to roam in portions of our landscape, even if the resulting ecosystems are more "cultural" than "natural," a dichotomy that I increasingly realize is both misleading and damaging.

26. Shovel-Tusked Gomphotheres Ate Bark

If I were a shovel-tusked gomphothere, like one of those that lived in our neighborhood between two and twelve million years ago, I wouldn't eat pine bark. Even if I were a mature male weighing five tons and supporting a prodigious set of tusks adapted for bark scraping, the little chips that I'd get off a big pine tree would be covered with resin, low in nutrients, and otherwise not worth the bother. No, if I were a shovel-tusked gomphothere, I'd head right for the maples and birches – if there were any around here back then – or at least go for the thinner-barked oaks.

The islands in the Miocene sea that are now connected and constitute peninsular Florida were even more recently home to bark-eating animals much larger than shovel tuskers, including mastodons and mammoths, but I'm still partial to gomphotheres. All of these animals are long gone, some at the hands of humans that wandered into our neighborhood from Asia only about 12,000 years ago. But even if this megafauna is now locally extinct, the tree species they ate are still around and available for study. I can't help wondering if our trees don't retain traits that helped them cope with the likes of the eight-foot-tall beavers or the even-larger giant ground sloths that now are only found in natural history museums. Sure, white-tailed deer and marsh rabbits eat bark too, and, more recently, we've witnessed what bark beetles can do to pines, but it's the really big bark eaters that have always captured my fancy.

Bark isn't very nice stuff to eat, unless the only alternative is wood. Even beavers and bark beetles know that it's only the inner layer of the inner bark – cells in and very near the vascular cambium – that has the soft-walled tissues with the most food value. Trees thicken by cell division around this trunk-encircling cambium and go to great lengths to protect it. It's no surprise that bark is packed with toxic, abrasive, and just plain tough tissues.

"Bark" is a word we use with caution around our house because my wife is among the world's foremost bark ecologists. She gets annoyed when we are not careful to specify whether we are referring to outer bark (the dead tissues produced by the cork cambium) or inner bark (known as phloem and composed of the living products of cell divisions in the vascular cambium). The vascular cambium also makes wood towards the inside of the stem. In most tree species, outer bark has solely protective functions – no fluids move through its cork-impregnated cells. Inner bark, in contrast, performs both protective and conductive functions – sugar produced by leaves, stored in roots, and allocated for fruit production mostly moves through living tubes in the inner bark. These tubes are often embedded in bands of tough-walled fibers that provide protection from aphids and larger sugar suckers.

Many tree species are readily identified by the appearance of their outer bark. The longitudinal fissures that typify ash bark, the bark plates that slough off pine boles, and the strips that peel from cypress trunks are all familiar features of outer bark. Thick layers of empty cork cells in the outer bark of longleaf pine render the large trees immune to fire damage, unless sloughed bark has accumulated around the base of the bole and is allowed to smolder for hours, thus heat girdling the trunk. The cork of commerce is the outer bark of Mediterranean cork oaks. As winebottle stoppers, they are unrivaled because their impregnation with suberin renders them quite impermeable.

In addition to some noteworthy physical properties, the chemistry of bark renders it tough stuff to eat, decompose, or burn. For example, some barks contain amazingly high concentrations of tannins, so high that mangrove, hemlock, and oak bark were traditionally used to tan leather. These same tannins confer some of the desired astringency to the red wines they confine to bottles. Then there are the barks that contain enough silica to dull saw blades, wear down beetle mandibles, and abrade beaver teeth. The barks of other species are used to produce various dyes, medicines, and poisons. All these features are produced and used by the tree just to protect the underlying cambium from damage.

Given that the outside of tree trunks must accommodate both growth in girth and wind-induced trunk sway, the toughness of bark needs to be of the flexible variety. By interspersing thick-walled fibers in softer tissues, bark remains pliable. In our area, people formerly took advantage of this pliability by fashioning cloth from the barks of mulberry and basswood trees – the interwoven fibers of their inner bark are durable, if a bit scratchy.

But does all of this bark biology reflect the activities of large and now-extinct mammals? I think it may, given that our familiar tree species lived for many millions of years with shovel-tusked gomphotheres and others of its ilk. In contrast, the 12,000 years that our pines and oaks have had to contend with humans has hardly been enough time for dramatic coevolution in bark properties. Even fire, which we consider a "natural" process, might not have been so important back when giant mammals were stomping around, knocking down trees, and eating everything in sight. But which of the properties of bark that we interpret as adaptations to fire, beetles, or beavers were actually the result of thousands of generations of trees having to contend with the dietary preferences and foraging strategies of shovel-tusked gomphotheres? Could the season of active cambial growth, during which bark is easily stripped, be the result of natural selection for avoidance of giant ground sloths? Might the tendency for some barks to come off in chunks rather than strips represent a mechanism to dissuade glyptodonts (picture armadillos on steroids) from bark munching? Did interlocking inner-bark fibers evolve in response to tusk-wielding proboscidians with a taste for cambia?

With a Bobcat, small bulldozer, or even a tractor with functional hydraulics, I could carry out tests of these and other hypotheses about coevolution of tree bark and gomphotheres. I'd weld up a set of tusks, attach them to the dozer, wire it up with force transducers and data loggers, and venture forth to do Science. Another approach would be to resuscitate large and now extinct beasts themselves by patching together the fragments of DNA found in fossils with the ancient sequences preserved in their living descendants. I've read a bunch about molecular genetics, and it doesn't sound all that hard. I suspect that if I were willing to spend a lot of time in the lab, I could probably pull off a gomphotheric reconstruction, with some help. But until a donor steps forward and I get tired of playing in the woods, I'll be content speculating about how life was back when beasts were big and humans were small, scared, or not yet in the picture.

27. Yaupon Redeemed

"What's in a name? That which we call a rose
By any other name would smell as sweet."

Act II, Scene II, *Romeo and Juliet*, William Shakespeare

In my ethnobotanical novella entitled "Timucuan Tea," the twelve-year old protagonist is accused of providing his school's football team with a controlled substance that miraculously converts them from losers into champions. Just as his fate seems sealed by the combined forces of the DEA and the irate coach of the defeated team, his father, a shabbily dressed but brilliant biochemist, saves the day. The father reports that his son's secret brew is just yaupon tea. He also reveals that the tea contains unheralded concentrations of antioxidants, lots of the antitumor agent ursolic acid, and a big dose of the methylxanthine alkaloids caffeine and theobromine. With those revelations, the boy looks more like a teenaged hero than a villain. Those revelations also make some entrepreneurial representatives of the local university worry about intellectual property rights and royalty sharing if they incorporate yaupon extract in their already-famous CrocJuice Energy Drink.

Yaupon is a common understory shrub that covers much of the Southeastern Coastal Plain, sometimes growing in densities that are problematic for plantation foresters who treat it as a weed. It's also a familiar yard hedge species, with a weeping form ("Pendula"), a small-leaved form ("Nana"), and dozens of other named horticultural varieties. On the UF campus and around Gainesville, it is the most commonly planted shrub. Pounds of leaves can be hand-stripped from the branches of planted or wild-grown shrubs in a matter of minutes. Yaupon holly leaves brew up into a tasty cocktail of stimulating alkaloids and health-conferring antioxidants. Yaupon tea was a daily pleasure for the Timucuan Indians of North-Central Florida and the Seminoles and Crackers who followed them.

The claim that yaupon tea is tasty was recently supported by the results of a blind taste test conducted by University of Florida (UF) undergraduate Alisha Wainwright. Alisha compared a tea she brewed from yaupon leaves with yerba mate, a tea made from the leaves of a closely related but commercially available *Ilex* species from South America. To her surprise, most subjects, including confirmed yerba mate drinkers, preferred yaupon. This result is compelling, given that yerba mate is exported from Argentina and Paraguay by the hundreds of tons each year, whereas yaupon is currently underappreciated as a beverage.

I can no longer hold off acknowledging that the scientific name for yaupon is *Ilex vomitoria* and that it was indeed used by Amerindians in ritual purification ceremonies. With these unfortunate facts revealed, I need to stress that biochemical analysis of yaupon foliage shows that it is no more emetic than coffee, tea, or any other caffeinated beverage. I suspect that this nomenclatural affront was commissioned by Ceylon tea merchants in England who wanted to crush competition from this native American product.

Early English chroniclers of life in our region were apparently more fascinated by ritual vomiting than about the sterling qualities of yaupon as beverage. Perhaps I reveal my Celtic roots, but I suspect those rich Anglo-Saxons were a prim bunch of pansies. There were indeed special occasions when Timucuan and later Seminole warriors vomited after drinking huge quantities of an especially strong brew of yaupon, referred to as "black drink," but that was only after fasting for days and singing, dancing, and generally carrying on for many nights – Kool-Aid would have had the same effect. And what's the big deal about vomiting, anyway? As a child, the mere mention of Sunday school or lima bean consumption could induce that response. A few minutes after my performance, I would be ready to go out and play baseball or eat ice cream, options seldom granted by my not-so-easily-fooled parents. I should add that ritualistic vomiting is practiced in cultures all over the world – even dogs recognize the benefits of an occasional purge.

The people first encountered by Spaniards in Florida exuded good health and towered over the newcomers by nearly a foot. The Timucuans of North-Central Florida, in particular, impressed those gold-crazed, disease-ridden, Inquisition-fleeing, and otherwise tormented Europeans by their size and grace. Timucuans did enjoy more protein than the bread-eaters from the Old World, but a heretofore unrecognized tradition of the Timucuans and many of their trading partners all over eastern North America was daily consumption of antioxidant-packed yaupon tea.

Much of what we know about the chemistry of yaupon tea is based on the research of Matt Palumbo, a UF graduate student who worked in Professor Steve Talcott's lab in UF's Department of Food Science and Human Nutrition. Using Steve's Waters 2695 Alliance HPLC System with a Supercosil LC-18 Column and PDA Detector (elaborate machines that work in, what are to me, mysterious ways to generate data that are interpretable only by biochemists), Matt discovered that, in addition to caffeine, theobromine, and other alkaloids, yaupon contains high concentrations of antioxidants, including a cocktail of flavonoids and various isomers of chlorogenic acid and coumaric acid. Concentrations of these compounds are highest in plants grown in full sun, whereas psychoactive alkaloid concentrations are especially high in young leaves from female plants fertilized with nitrogen. Females can be distinguished by their stamen-lacking flowers or by the presence of little, round, red, juicy fruits that are favored by brown thrashers and bluebirds. Matt also found that leaves of the Pendula variety contain higher concentrations of caffeine than the Nana variety, but that nitrogen fertilization increases caffeine severalfold in both wild and cultivated plants.

Steve Talcott, now at Texas A&M University, has continued to explore the chemistry of yaupon, especially its exceptionally high concentrations of antioxidants. In a recent study from his lab, for example, yaupon extracts were shown to have anti-inflammatory and various other chemo-preventive effects.

As pointed out by C.M. Hudson in his classic (and recently reprinted) book *Black Drink*, yaupon tea was widely marketed through the 18th and early 19th centuries in the Carolinas as cassina, in England as Carolina tea, and in France as Appalachina. Use of the vernacular name cassina caused some confusion about the tea source because *Ilex cassine* (dahoon holly) also grows on the Southeastern Coastal Plain. That confusion was cleared up by research in botanist Brad Bennett's lab at Florida International University. Based on its high caffeine concentration and high caffeine-theobromine ratio, the researchers concluded that *Ilex vomitoria* was the likely source of the tea that enjoyed such widespread favor. Those vibrant local and international markets for yaupon crashed after the Civil War at least partially due to its association with indigenous people and the poor. The scientific name conferred on yaupon in 1789 by Scottish botanist William Aiton certainly did not help its reputation. Since then, it hasn't been widely used locally or internationally.

While ceremonial uses of the black drink are described *ad nauseum* in even the scientific literature, milder brews of yaupon were widely consumed as a daily beverage by settlers and Amerindians alike. I strongly suspect that during the Seminole Indian Wars, the effectiveness of native warriors was enhanced by yaupon consumption. Note that the "asi" in "cassina" is the Muskogee name for *I. vomitoria*. "Asi-yahola," which means "Black Drink Singer," comes to us as the familiar name "Osceola," the famous Seminole leader. Hundreds of years before those bloody wars and hundreds of miles north of the species' natural range, other Amerindians were also enjoying yaupon tea. This evidence for an extensive trade network in North America more than 500 years before the European onslaught was only recently discovered by a team of chemically savvy archeologists excavating in Greater Cahokia, the massive pre-Columbian ruins near current-day St. Louis.

To make a quick cup of yaupon tea, pan roast a handful of fresh leaves until some turn black, then crunch up the crispy leaves and brew them as you would any other tea but be liberal with the leaves. If you have more time, air-dry the leaves for a few days and then dry them thoroughly in a warm oven. Timucuans reportedly boiled their teas for hours, but this might have only been for ritual cleansing with black drink. For daily consumption, they apparently preferred "white drink," a weaker brew with a froth raised by blowing into it through a hollow reed – Timucuan cappuccino. Paraguayans, on the other hand, cool-smoke their yerba mate leaves before crushing and brewing.

Unfortunately, given the ritual captured by yaupon's scientific name, I was worried that the species will continue to be used primarily as an ornamental shrub in suburban gardens and sprayed with herbicides in pine plantations where it grows wild. Although Alisha's study showed that, on the basis of taste, even yerba mate drinkers preferred yaupon, she also found that knowledge of yaupon's scientific name would keep most people from purchasing it. In other words, the Shakespearean quote with which this article commenced apparently does not apply to beverages. This finding is unfortunate, given that, for millions of people who live within its native range, yaupon could provide a local, caffeinated, antioxidant-rich drink, one that offers an otherwise-healthful alternative to imported tea, coffee, and other caffeine crops.

My concerns about the potential for a revival of yaupon tea recently diminished when I realized that no less than four companies are now marketing it. This is great news, but what will happen if yaupon becomes an overnight market sensation? Will an exponential rise in demand for yaupon tea result in excessive resource mining followed by domination of the trade by agribusiness? Will consumers purchase yaupon tea made from genetically engineered clones grown in noxious chemical slurries of inorganic nutrients? I am actually not too worried. Given the ease with which yaupon is cultivated as a hedge, I envision neighbors joining together to grow their own leaves in the shade of restored pine savannas. And, rather than roller-chopping and herbiciding yaupon, plantation foresters might begin to treat it as a valuable non-timber forest product, which will have numerous environmental and financial benefits.

28. I Saw a Sandy-Mounder

The burrowing rodent *Geomys pinetus* goes by many names including "pocket gopher," "sandy-mounder," and even "salamander." My son calls the sandy-mounder that frequents our persimmon orchard "Clara," but I have no idea how he determined its sex or how he knows there is only one beast tunneling its way among the persimmon, pomegranate, and fig trees.

There's justifiable confusion over what sort of creature makes the ankle-high earthen mounds that pop up in fields, gardens, and pastures. Gopher tortoises pile soil around the aprons of their burrows, but those excavations are easy to distinguish by the presence of a large oval entrance hole. Fire ants also pile fresh-dug soil, but pain is a good teacher, so most people quickly learn to give their nests wide berth. There's even a ground-nesting scarab beetle (*Peltotrupes youngi*) that makes a small pile of soil in the process of excavating a yard-deep narrow shaft that leads to a chamber lined with leaf litter onto which the females lay their eggs. But it's lines of large piles of fresh sand at ten-to-twenty foot intervals that is a sure sign of a sandy-mounder at work.

The other evening I was out in the field behind our house, escaping domestic duties, when a puff of sand came out of the ground not far from where I was reclining. I sat up a bit to get a better view and, a few flicks of sand later, the front half of the excavator emerged. It was a small, roundish mammal, kitten-sized with soft-looking fur of a light agouti brown – almost tan. A rare sighting of the seldom-seen pocket gopher! Its cute face seemed to have a furrowed brow of intense concentration, but that impression might have been because it was covered with sand.

Pocket gophers were featured in some of my favorite childhood cartoons – usually playing tug-of-war over a carrot with some hapless farmer – so I was quick to assign the little beast I was watching some pretty complex and appealing personality traits. Whether or not this one was as playful as I imagined, it had made a knee-high mound of sand in the space of a few minutes. Judging by the color and moistness of those tailings, they must have come from six inches to a foot below the surface. Then the beast was gone, presumably scurrying along its extensive system of tunnels looking for something to eat. I see sandy-mounders about once per decade, so this sighting was a treat.

Pocket gophers make mounds of sand in the process of tunnel excavation. Those mounds represent only a fraction of the estimated ten cubic yards of soil they move per-acre per-year; the rest is stuffed into old tunnels as they open new ones. Remarkably, a single half-pound animal usually maintains about 200 feet of tunnels. Given that mound and beast size are closely correlated, you can tell if your beasts are juveniles or adults. But sometimes it's hard to determine how many of these entirely fossorial animals are at work in an area. They are ferociously territorial, but tunnel systems can interdigitate with separations of only a few yards.

Pocket gophers make tunnels to gain access to the roots, which are their staple food of choice. For whiling away the winter and for raising young, a sandy-mounder mom will also open a chamber 3-4 feet below the surface that is accessed by a spiral ramp.

Unlike moles, which push up soil along the length of their near-surface tunnels, sandy-mounder presence is made evident by the widely-spaced spoil heaps. Another difference is that, whereas moles are carnivores that relish worms and beetle grubs, sandy-mounders are root-feeding herbivores. In both cases, given that digging through soil requires several thousand times more energy than moving the same distance above ground, both moles and pocket gophers have voracious appetites.

The appetites of pocket gophers are satisfied entirely by vegetative matter. They mostly eat roots, but sometimes they pull entire plants down into their burrows. Several published estimates of just how much they eat hover around 25 percent of total belowground growth. Not only that, but they are somewhat fussy eaters, disdaining fibrous grass roots for the more succulent roots, rhizomes, and tubers of broad-leaved plants. In addition to turning over the soil through their excavations, these little ecosystem engineers may thus determine what species of plants grow where. Tree and shrub roots are not immune to their browsing, and there are good reasons to think that pocket gophers help keep our invading pine savannas from being choked out by woody plants.

Creatures like our pocket gophers are found in grasslands and savannas around the world. Among these often only distantly related ecological cognates are mole rats and spring hares in Africa; hairy-nosed wombats and burrowing bettongs in Australia; and vizcachas, maras, and degus of South America. The mammalated surfaces these fossorial beasts create are often rich in annual plant species that benefit from soil disturbances. In contrast, I have not observed any distinctive plant species associated with the burrows of our native pocket gopher, but our severely drained soils and intermittent rains make establishment from seed a very chancy affair, and not all that common.

Sandy-mounders only forego their solitary subterranean existence briefly during the mating season. When in the mood, they emerge and find a tunnel of a sandy-mounder of the opposite sex into which they rapidly burrow. Given this lifestyle, it's not surprising that few people have seen one. Having been so blessed by a view of a sandy-mounder at work, I waited for a while to see if it would re-emerge. Finally, it got pretty dark, so I went back inside to find the table cleared, the dishes washed, and the children tucked in their beds, waiting for nighttime stories. I told them about pocket gophers until they fell asleep.

29. Moss Picker

It's been several years since I last saw the old man harvesting Spanish moss from the live oaks behind Prairie View School. Whenever he was there, standing on the hood or roof of his rattletrap pickup, I'd park my bike, jump the fence, and go over to visit for a while. No matter what sort of frantic day I'd had, there was timelessness about him working with his long, nail-studded pole up in those moss-festooned crowns that calmed me. I watched in fascination as he reached up into the long dangling clumps of Spanish moss and snagged them with a twist of his trusty pole. Although he seemed to move in slow motion, the ground around his truck was quickly covered with moss, which he would occasionally pile in the truck bed. One time, I arrived as he was driving off with a full load, the sale of which would just about pay for his gas, though probably not for the oil his truck obviously burned in abundance.

Our somewhat formal relationship was comfortably based on what I imagine are established local protocols for how ancient black men should deal with youngish white ones. I learned from him, and I probably confirmed some of his suspicions about guys like me, though I believe he still appreciated my interest.

"Green" moss, like the stuff the old man sold, is currently used fresh or dried in floral arrangements. Traditionally, a higher price was paid for black moss, the horsehair-like fibers that remain when the living tissues of the leaves are rotted away. Pickers sometimes rendered their own moss, but the composting process takes months, and it was more often done at a moss ginnery; the last in our region burned down in 1963. At one time, black moss was the principal filler for mattresses as well as overstuffed chairs and sofas. I thought that black-moss upholstery went out of style around when men stopped greasing their hair back with macassar oil, but it was the stuffing in the bucket seats of my 1971 Opel.

Partially because of its vernacular name, Spanish moss is a most misunderstood species. First of all, it's a native of America, not Spain. Interestingly, its species epithet of *usneoides* refers to its resemblance to old man's beard lichens, to which it is also not evolutionarily related. Second, it's a flowering plant and not a moss. Third, it's a rootless epiphyte, not a parasite; it gets all its water and nutrients from rain, mist, and dust while it takes nothing from its host tree other than support.

Tillandsia usneoides, our familiar Spanish moss, makes for durable upholstery, but, as a living plant, it's nothing short of amazing. How does it survive up there in the treetops without functional roots? Keys to its success as an epiphyte are the tiny flat-topped hairs that cover its leaves. Each hair, which is composed of eight cells, has a stalk that penetrates down into the leaf surface with a large plate-shaped cell capping the top. Imagine a plate held up on a pedestal. The plate analogy is apt when the plant is wet and the leaves look green. Dry Spanish moss leaves are whitish because the plate edges reflex upwards, converting the plate into a bowl. The white surfaces of the upturned cells reflect light, which helps to keep the leaf from heating up and thus reduces its need to evaporate water for cooling purposes. Upward flexure of the plate margins also pushes the stalk cells down, which causes them to expand and plug the gap that surrounds them when water is available.

When Spanish moss plants look green, they are absorbing water that moves by capillary action under the plate and down though the space around the stalk into the leaf interior. Water uptake is precluded, but water loss rates are greatly reduced when the plates turn into reflective bowls. For a rootless plant up in the canopy, avoiding water loss is the principal challenge if it isn't raining or hasn't rained recently. The problem with the Spanish moss mechanism for avoiding water loss is that it also limits uptake of the carbon dioxide from the atmosphere it needs to photosynthesize. Spanish moss solved this dilemma in a way that eluded plant physiologists until a few decades ago.

Unlike most plants, it's only after the sun goes down and the humidity goes up that Spanish moss absorbs the carbon dioxide it uses for photosynthesis. Inside the leaf, that carbon dioxide gas is converted into an organic acid and stored in that form until the lights come on in the morning. When the solar energy needed to make sugar is available, the carbon dioxide is released from the acid, right where and when it's needed in the sugar-making process. If you're willing to chew on mouthfuls of Spanish moss, you'll find that its flavor in the morning is decidedly tangier than it is in the evening, by which time the sequestered acids have been spent on sugar. This pathway to photosynthesis, which physiologists call crassulacean acid metabolism, or CAM, isn't a good way to grow fast, but it serves to save water when it's scarce. No wonder that cacti, epiphytic orchids, and many species of the pineapple family – of which Spanish moss *is* a member – employ CAM.

Spanish moss may not superficially look much like a pineapple, but the two species are definitely related. Although pineapple fruits are succulent and Spanish moss fruits are dry capsules filled with wind-dispersed seeds, their flowers are structurally similar. Unlike the many flamboyantly flowered members of the pineapple family, Spanish moss flowers are small, green, and fairly inconspicuous, but they're still worth seeking out in March or April.

Although Spanish moss flowers and fruits abundantly, I'm not sure that it often propagates itself by seed. What happens a lot is that new patches are established from fragments broken off and blown by the wind. I learned, during a somewhat-controlled burn with one of my classes, that glowing clumps of Spanish moss also fly up in fire-induced updrafts, making it one of our most effective firebrands.

Spanish moss is found across the Southeastern Coastal Plain and in scattered populations down into Chile and Argentina. Locally, it proliferates on trees with moisture-retaining bark that doesn't tend to slough off in big plates, hence its preference for live oaks over pines. Trees that have leaves which readily leach nutrients tend to support bigger crops of Spanish moss than trees that better protect their potassium. Spanish moss also seems to proliferate on dying trees, perhaps because moribund trees lose their capacity to retain nutrients that the Spanish moss then uses.

I'm pained by billboards that advertise Spanish moss removal services. I generally hate billboards, but it's particularly distressing that some Southern homeowners pay to have their trees demossed. Some people harbor the misconception that Spanish moss hurts trees, but I've also heard that some people don't like its looks. For me, moss-draped live oaks are a Southern icon.

30. Guilt and the Cabbage Palm

To shade a window through which far too much sunshine pours during the summer months, we're considering the strategic planting of a large cabbage palm. For $200 or so, we can have a big palm dropped off next to a predug planting hole. If we can get our neighbors to share in a full truckload of palms, the price will drop. And if Richard gets personally involved and the hydraulics on his tractor cooperate, then the task of manipulating two-ton palms into planting holes and holding them upright while strapping on braces will be greatly facilitated.

While I'd like to look out the window at a large palm and would love to hear the fronds susurrating in the breeze, the idea of buying a tree that was yanked out of some forest for our horticultural benefit gives me pause. Research on cabbage palm growth rates causes me to be even more concerned: I know that a forest-grown palm with a twenty-foot trunk may be more than 200 years old. That extraordinary figure is derived from the number of leaves a cabbage palm in the wild produces annually – let's say three on average – with a per-leaf height increment of less than an inch. To this, you must add the length of time an average cabbage palm spends building a big stem belowground – about thirty-five years in the wild. This peculiar establishment-growth phase is necessary in palms because, once they start aboveground growth, their stems can no longer grow in diameter. Using these estimates, it's easy to see why I expect that the palm we're contemplating buying started to show an aboveground stem around the time Andrew "Sharp Knife" Jackson was slave-raiding Seminole villages in the new state of Florida. I'm not worried about our proposed transplant surviving, but, somehow, I feel some chagrin about messing with an organism of such an exalted age, especially if I thereby contribute to the extirpation of the species in the wild.

We could go for a smaller palm that is younger by perhaps a century, but transplantation success increases with size. This phenomenon seems peculiar to palms and is apparently related to their dependence on stem water storage when groundwater is limited. More to the point, a shorter palm won't shade our window until I'm too old to enjoy the benefit.

My concerns about the effects of our proposed palm purchase on wild populations of our state tree are somewhat mollified by its abundance. In fact, researchers and park managers are trying to figure out why cabbage palm populations seem to be expanding at the expense of both pines and hardwoods.

Ecologists are prone to invoke fire suppression whenever some population goes haywire in Florida, but this explanation doesn't work in the case of cabbage palms. When Kelly McPherson was studying the palm population explosion issue for her thesis project, she subjected large and small cabbage palms to both high and low intensities of controlled burns in summer and winter for two years, but these fires still did little permanent damage. Lacking the bark and sensitive cambial tissues that bark protects, big palm stems are unharmed by even the hottest fires. The terminal bud is buried deeply down among the leaf bases, where it, too, is well protected from fires. In fact, cabbage palm coronas burn spectacularly, often scorching back encroaching branches of other trees, thereby securing a place for themselves in the sun.

Although the terminal bud of a cabbage palm, called a palm cabbage, is extremely well protected, it's quite tasty. The edible portion is about the size of a parsnip, but, even with a sharp machete and a functioning chainsaw, I've never found palm cabbage harvesting to be worth the effort. Furthermore, extracting the cabbage kills the palm, which, in light of their often-advanced ages, seems pretty wasteful. Nevertheless, I've seen palm cabbage on menus a time or two, and I've ordered the dish once out of curiosity – it came with mayonnaise and mini-marshmallows, which just added insult to injury. Bears eat palm cabbages plain, but I can't believe that the reason there are now so many palms is because there are so few bears. Within fifty miles of Cedar Key, I suspect that palm populations are still recovering from the hammering they suffered while the Standard Manufacturing Company of Cedar Key was making Donax brushes using cabbage palm trunk fibers. During that period (1910-1952), many thousands of palms were chopped down and ferried out to the factory where seventy-five workers fashioned the fibers into clothes brushes and dust brooms.

Although there weren't many, if any, cabbage palms at Flamingo Hammock when we first bought the property, the few we planted thrived and reproduced like gangbusters. The transplants come rootless and with most of their fronds hacked off; it takes a year or two for their coronas to recover fully. Recently, I've seen a lot of young cabbage palms coming up, but I sometimes probably confuse them with the much-more-common bluestem palms, its diminutive relative.

One reason for the abundance of cabbage palm seedlings around our place is perhaps the local and probably temporary scarcity of palm weevils (*Caryobruchus gleditsiae*). In natural cabbage palm stands, the larvae of this seed-eating beetle devours most of the palm seeds that the mice miss, especially those that fall abundantly near fruiting trees or that are concentrated in bear and raccoon scat. A gravid female beetle glues a little white egg on the outside of a seed or fruit. The larva soon hatches, burrows inside, and consumes the embryo and its food stores. About thirty days later, an adult beetle emerges and the cycle begins again. Only seeds that are dispersed away from mother trees and deposited singly are likely to escape these little predators. Given the gastrointestinal effects of eating too many palm fruits coupled with the natural gluttony of bears, raccoons, and their kin, many cabbage palm seeds are indeed well dispersed and not concentrated in discrete piles. I am pretty sure that this capacity to cause diarrhea is the product of the cabbage palm's natural selection process on its fruit's chemistry.

Given that cabbage palms actually seem to be thriving in the wild, I've convinced myself to plant a big one, perhaps even two. With such a high likelihood of survival, and with the importance of palms to wildlife, it seems like a worthwhile investment. What I won't do is import a palm from Hillsborough County, where cabbage palms are dying by the thousands due to a microscopic pathogen called a micoplasm. This disease threatens to be as bad as Lethal Yellowing, the related disease that decimated populations of several palm species in Southern Florida.

31. Chiggers, Ticks, and Pesticide Drift

On the deep sands on the west side of Gainesville is an upscale, golf-course-centered development that has received several environmental awards, apparently for leaving trees along fairways and roads. Unfortunately, in the absence of fires, hardwoods are moving in and pines and flowering herbaceous plants are moving out.

Mostly due to the persistence and persuasiveness of Kim Heuberger, a graduate student at UF, we received permission to reintroduce fire into a dozen patches of remnant sandhill in that posh subdivision. The patches we burned bordered fairways, constituting what golfers call the "rough." Because fairway margins are also prime building lots, the strips of rough often back up on expensive houses. To keep the rough from getting too rough, or at least too thick, the golf course manager regularly sent out crews with chainsaws, weed whackers, and herbicides to beat down oaks and other hardwoods. He liked our suggestion of testing controlled burns as an alternative vegetation-management technique.

Kim is the sort of modern ecologist who is responsible and thinks about the social and political dimensions of research projects. She wisely selected Allyson Flournoy, a respected environmental law professor, and Nancy Pywell, an environmental education specialist, as the other members of her advisory committee. I sealed my reputation during Kim's first committee meeting when, in response to a query about equipment needs for the project, I quipped something about matches and beer. But Kim did it right. She displayed posters describing the project, leafleted neighbors, conducted neighborhood meetings, and assessed attitudes and knowledge of relevant stakeholders before and after the controlled burns. Her Nomex-garbed burn crew was efficient and professional.

Lest you imagine raging conflagrations, let me point out that some of the burn sites were only slightly larger than the swimming pools they bordered. Furthermore, fires had been suppressed for so long that there wasn't much in the understory to burn other than oak leaves. In fact, besides a few scraggily clumps of wiregrass and some scattered elephant's foot, there wasn't much evidence of what had been there before development resulted in fire suppression: a diverse array of wildflowers. No wonder! The canopy was choked with hardwoods, and the remnant longleafs were barely keeping their heads above the tide of invaders. We had to drip a lot of fuel to get the patches to burn, but, even then, the burns weren't hot or thorough. Nevertheless, they were apparently much enjoyed by the mostly retired neighbors who came out to watch, rake a bit, and of course give advice and tell stories of their own experiences with fire.

Although the controlled burns weren't spectacular, they produced some of the hoped-for results. The golf course manager was happy that most of the small oaks were top-killed. He explained that stand opening increases airflow over the greens and fairways, which reduces their need to apply fungicides to keep the sensitive grasses healthy. As expected, even the single burn increased the number and diversity of plants in flower. Reportedly, the blazing stars and other showy wildflowers were appreciated by people golf carting past the burn patches. During an informal meeting at what is familiarly known as the 19th hole, several golfers reported that the burning helped reduce their scores, which Kim and I both found baffling.

Later, one of the golfers explained that there is a two shot penalty for lost balls, and their white balls were easy to find on the black stubble.

In response to Kim's energetic environmental education campaign, and having seen the effects of the fires, the community at large showed vastly increased knowledge and acceptance of controlled burns as an ecosystem management tool, even in their suburban setting.

One component of the project that didn't pan out was the reduction of tick and chigger populations by fire. Fire ecologists and piney-woods burners of other persuasions agree that these pests are greatly diminished after burns. For this part of the study, we had planned to estimate tick and chigger populations prior to the fire and then, afterwards, assess populations at fortnightly intervals using standardized protocols. Tick populations are estimated by counting the little beasts on a yard-square sheet of white blanket material after dragging it ten yards across the ground at a walking speed. Population densities of chiggers, also known as "red bugs" – though they are actually the larvae of mites – are estimated by placing a black vinyl LP (33 1/3) on the ground in the mid-afternoon and counting the chiggers on the surface after one hour. Apparently, hot LPs outgas some compound that chiggers can't resist. To reduce bias, Kim selected LPs by male country singers. We conducted a test of both techniques in the field in front of my house. The results were enough to make me want an office job.

To our surprise and in startling contrast to what we found at Flamingo Hammock, we found neither ticks nor chiggers at the golf course study sites even before the burns. This was bad news for that part of the study, but it seemed like good news for the poor beasts upon which ticks and chiggers feed. In any event, we were too busy with the burns to think much about the lack of ticks and chiggers. Only later did we wonder at just how much pesticide drift it took to rid an area of these hardy pests. Had I ever dreamed of a fairway home in an upscale golfing community, that dream would have died the day this part of our experiment failed.

32. Southern Pine Beetles

If we delay thinning our loblolly pines much longer, we're likely to lose them all when the next plague of southern pine beetles (SPBs) washes over Alachua County. We lost a few acres of pines to the outbreak in 1998, and another bunch in 2002. If populations of the rice-grain-sized beasts continue on the same cycle, we're due for another attack, and the trees in the stands that had escaped last time are now large enough and crowded enough to be attractive to SPBs, especially if the rains start late or stop early. Under those conditions the loblollies would be under water stress, and they'll likely end up as beetle fodder. In other words, we're faced with either having a logger come in to take away two-thirds of the live and semi-marketable trees now, or waiting until after they are all dead and no longer have sale value because of the glut on the market of beetle-killed trees.

Although loblolly is still not my pine of choice, I'm starting to appreciate its virtues. Northwards, in the land of the Tar Heels, loblolly is a species as worthy of respect across the landscape, but here in Florida prior to agricultural land abandonment, it was a species found only on floodplains. Most of our loblollies are colonists of abandoned agricultural clearings and pastures. It's a good colonizer because it produces lots of cones with lots of seeds that are small and winged, so they're well equipped for wind dispersal. After establishing themselves on a fertile site, young loblolly trees can grow rapidly. When we bought Flamingo Hammock in 1984, the old pastures were covered with loblolly pines as thick as the fur on a dogs back, but thinning was the work of a well-swung bush ax. Toppling the mostly malformed and fusiform rust-infected trees often required only a single blow. Now the trees are large enough for sawtimber, or at least a chip-and-saw harvest.

Despite having run quite a few fires through the developing stands of loblolly pines, the trees are now close enough to compete for water. This competition will exacerbate the stress from the drought for which we are due. The stressed trees will then, for reasons that elude me, emit chemical distress signals that will be intercepted by gravid female SPBs. The invasion will then begin.

The first SPB colonists to land on an even badly stressed pine tree need help in chewing their way down through the outer bark to the sweet stuff beneath. To recruit this help, they emit their own chemical signal, called a pheromone, which attracts other SPBs. Working together in teams of three or four, they dig down through the bark to the bark-wood interface, where they excavate a chamber and lay some eggs. It's not clear to me whether the adult SPBs eat much of anything, but the larvae devour young bark and wood, along with the vascular cambium, the ever-young tissue that produces both. The resulting tunnel systems are winding and crisscrossing, which distinguishes them from the tunnels of engraver and turpentine beetles, whose tunnels tend to be more linear and oriented parallel with the grain.

A healthy loblolly pine, unstressed by neighbors or drought, can "pitch out" SPBs by exuding resins under sufficient enough pressure that the gnawing adult females will fail to reach the cambium. In contrast, when a pine is stressed, its resin pressure goes down and so do its defenses. Once a tree has several thousand beetle larvae munching around under its bark, there's not much that can be done to save it. First the tree's needles turn red, then brown, and then the tree is quite dead. Long before the needles brown, the mature beetles have already hatched and flown off in ever-increasing numbers to find other trees to colonize.

SPBs are weak fliers, so pines separated from conspecific neighbors by thirty feet or more often escape. Unfortunately, the crowns of our loblollies are nearly touching, which means that, to reduce the SPB risk, a lot of them have to go.

As part of our pine savanna restoration efforts, we've slowly been replacing the loblollies in our upland areas with longleaf pines. It probably would have been faster to have cleared all the loblollies at the get go, planting longleaf pine seedlings in the resulting clearings, but no one wanted to look at a big clearcut. Furthermore, fallen loblolly needles carry fires well, which helps a great deal with running controlled burns through the soon-to-be longleaf pine savanna.

It appears that SPB outbreaks can be contained, but only with concerted and coordinated effort. Although long-distance dispersal by SPBs does occur, if the red tops of attacked trees are spotted early enough and all the loblollies within 100 feet are cut and removed, then the remaining trees are generally safe. This strategy seemed straightforward enough, or so thought my friend Meg, the City of Gainesville Arborist during the SPB outbreaks. Her approach was to cut down infected trees before they released their SPBs, thereby reducing the threat to many other trees, loblollies and longleafs alike.

Hiring a professional, bonded, and insured tree surgeon to remove live loblollies ain't cheap, but dealing with dead trees is more dangerous, and the cost consequently spikes up. If you have enough live-but-threatened loblolly pines, then they might even be harvested at a profit. What neither Meg nor anyone else on the SPB Advisory Board anticipated was the anti-SPB control broadside launched by fellow environmentalists. We did expect that some residents would be loath to part with the cash needed to have some trees removed to save others, especially if the trees saved were not on their property. What we didn't expect was that some of our colleagues in the environmental management field would argue that the SPB outbreak should be allowed to run its course without our interference.

Critics of the SPB containment campaign argued that the SPB is a native species, not some recently introduced pest. Despite the lack of records of SPB outbreaks in our area, they regularly decimate pines up in the Carolinas and further west. The opponents of SPB containment efforts also reminded us of the colossal, expensive, and environmentally destructive efforts to control outbreaks of gypsy moths and spruce budworms in the North. Finally, they pointed out that most of the loblolly pines in our area are the result of agricultural land abandonment, not fully natural processes.

But there is more to this story. It would've been tragic enough if all SPBs ate was loblolly pines, but when loblollies become scarce, SPBs actually turn to other pine species. SPBs apparently find spruce pines delectable but, when hungry enough, they move down the menu to slash and finally longleaf pines. I find it especially tragic when unchecked populations of loblolly-pine-fattened SPBs go on to kill our precious and otherwise-beleaguered longleafs.

The debate between supporters and critics of SPB containment fizzled out when SPBs started to attack slash pines. After that, support from forest industries was easily forthcoming. Then, when SPBs started to kill lots of venerable, old longleaf pine trees in San Felasco State Preserve, even the most outspoken proponents of the "let nature take its course" philosophy had to reconsider their position.

Gainesville's SPB controversy is good preparation for future threats to resources valued by society for which society-wide support for management responses will be required. Unfortunately, few people will bemoan the loss of our four species of ash if current efforts at containing the ash borer fail, but I'll miss their brilliant fall colors. That red bay, swamp bay, and sassafras have been wiped out across the South by the one-two punch of an introduced beetle and its hyper-virulent fungal associates is hardly on many radar screens. Perhaps public response will be stronger if the "sudden oak death" that lurks over the mountains in California or the "oak wilt" in our backyard in Texas comes knocking at our door? Can you imagine life in Gainesville without live oaks? Even if there is someone out there who doesn't particularly care for oaks, the loss of oak shade will increase our collective heating, and air conditioning demands will skyrocket and so, then, will utility bills.

For now, vigilance is what is needed in the face of ash borers and threats to our live oaks. In contrast, SPBs are with us, and we need to assure that the environmental, fiscal, and horticultural disasters of the recent past are not repeated. For my part, it's about time I got around to thinning our pines.

33. Cypress Knees: Root Snorkels or Dinosaur Defenses?

Slogging through cypress swamps is great fun until you bark your shin on a cypress knee. Despite the occasional bruise, I share (with bric-a-brac collectors, wood carvers, and traditional fishnet float users) a fascination with these strange protuberances.

Cypress knees are known by scientists as pneumatophores (Latin for "breathing roots"), despite the lack of evidence for this function. Given that the swampy soils in which cypress trees root are low in oxygen, it would seem reasonable for them to put up snorkels for a breath of fresh air. Unfortunately, when a scientist covered cypress knees with plastic and measured oxygen concentrations in the roots, he found no effect. Given the results of this experiment, I favor Walter Judd's explanation that the pointed knees helped protect the trees from web-footed dinosaurs, intent on munching on cypress needles. Walter is a widely respected botanist with a lively imagination.

One hundred million years ago, when dinosaurs were dominant and our furry ancestors were scarce, small, and scared, cypress trees had to deal with the big reptiles. Similarly, only ten thousand years ago, long after the dinosaurs took to the skies or went extinct, Floridian mastodons also fancied cypress needles (judging from studies of their gut contents by paleontologists from the Florida Museum of Natural History). Given that bald cypress trees can live to be more than one thousand years old, ten thousand years is only a few tree generations ago for them.

It doesn't matter much to me whether dinosaurs or mastodons had a bigger influence on the evolution of this ancient species; I am just enthralled by the challenge of incorporating such historical factors into interpretations of the current biology of cypress. Walter did so by envisioning cypress knees as defenses against web-footed dinosaurs and other large beasts – barking your shin on a cypress knee is unpleasant, but stepping on one might be a serious annoyance, even for a beast of several tons.

Even without their dinosaur-repelling pneumatophores, cypress trees call out for attention. Consider a deciduous conifer – a pine relative that drops its needles together with the small twigs to which they are attached. Being leafless during the winter isn't hard to understand, in Illinois, which is the northern end of the range of this species. But down in subtropical southern Florida, it is surprising that cypress trees spend leafless months while their evergreen neighbors soak up winter sunshine. Once again, an explanation for this apparently paradoxical behavior can be found in the species' history.

For most of the one-hundred-million-year history of bald cypress and across most of its once-vast historical range, Earth was warmer than it is today. In fact, the past two million years – the Ice Ages – have been uncharacteristically harsh. But, even though the world was warmer, it was still dark for half of the year at high latitudes. Perhaps cypress trees are winter deciduous because, when the species evolved in the then-warm far north, there was too little solar radiation for half the year to warrant leaf retention.

Although their exposed knees draw stares, cypress wood also deserves consideration. Back when houses had wooden foundations and fence posts weren't impregnated with poisonous preservatives, cypress heartwood was in great demand because it is so incredibly decay-resistant. Unfortunately, most of the cypress that is on the market today is sapwood from younger, faster-growing trees, and doesn't resist decay much better than pine.

The only really nice cypress lumber I've seen of late was "deadhead" logged off river and lake bottoms, where it had sat for a century or more. Submerged down in the goopy and oxygen-free muck, the admirable qualities of the sunken logs were preserved and perhaps enhanced. The wood sure is pretty, but deadhead logging causes a lot of damage to bottom sediments and the organisms they contain. These sediments also preserve important archaeological artifacts, such as the 2,000-5,000-year-old dugout canoes that emerged from the muck when Lake Pithlachocco (a.k.a. Newnans Lake) went dry during the 1998-2002 drought. Most of the one hundred canoes were pine, but some were hollowed out of cypress logs.

The cypress trees that yielded the logs that were converted into the canoes that plied the waters of Pithlachocco must have resembled those that survive on a short stretch of shoreline where old growth stands. For some reason, when the rest of the cypress trees bordering Pithlachocco were logged and rafted down to the steam-powered sawmill at the mouth of Prairie Creek, this patch of a few acres was spared. I'm not sure how old the trees are, but the largest are more than three feet in diameter and 100 feet tall.

After a recent spate of hurricanes, I was impressed with how well the old cypress trees weathered the storms. While many sweetgum and red maple trees were completely trashed, the cypress trees were apparently unscathed. Whether or not cypress root systems breathe through their knees, they are surprisingly wind-firm. Its small crowns and sparse foliage, even in mid-growing season, would also help the trees avoid being badly buffeted. Finally, the rot-resistance of cypress wood must also come in handy during the hurricane season.

Although several local governments in Florida and the Florida Department of Transportation prohibit the use of cypress mulch on environmental grounds, the process of cutting down and chopping up entire cypress swamps continues. Cypress could be sustainably managed for timber in swamplands throughout the South, but it isn't. There are no management plans, no silvicultural treatments to promote regeneration or stimulate growth, nothing but timber mining. It's incredible that it's still legal in Florida to clearcut cypress domes. Given that cypress does not recover in those clearcuts, buying cypress mulch contributes to ecosystem destruction.

This sad state of affairs is unfortunate because cypress-dominated ecosystems provide important habitats for storks, limpkins, and a variety of other wildlife. Cypress swamps are also like big sponges that absorb and filter water, releasing it slowly during dry periods. These sponge-like qualities make cypress swamps ideal for treating sewage, a well-researched but as yet underutilized environmental service, provided for free. Overall, I'd say that being chopped up for yard mulch is a sad fate for such an ancient and otherwise venerable species that has gone through a lot over the past one-hundred-million years.

34. In Praise of Old-fields

A portion of the largest natural area on the UF campus was recently plowed. The shrubs and young loblolly pine, red maple, and sweetgum trees were knocked down and turned under, which left bare mineral soil exposed on the surface and set the "successional clock" all the way back. If the area had been left alone, which might seem reasonable for a natural area, succession would have proceeded, the tree canopy would have soon closed, and a forest community would have developed. Instead, the newly plowed land is now open for colonization by ragweed, broomsedge, dogfennel, goldenrods, rattlepod, and a variety of other herbaceous old-field species.

The primary justification for mowing down and plowing up a recovering patch of forest is that one of the habitats most rapidly disappearing in North Florida is abandoned agricultural fields. The reason for this loss is that less and less area is being cultivated for vegetables, and, when a field is abandoned, housing developments or shopping centers usually spring up instead of goldenrods. The other reason why patches of UF's Natural Area and Teaching Laboratory (NATL) are plowed at regular intervals is that Tom Walker, the person most responsible for its existence, is an entomologist.

Old-fields are great places for flowering plants and wildlife, particularly insects and other arthropods. If you want to watch pollinators in action, hope to see an act of predation, or want to do an experiment on herbivory, avoid forests and find a field. In the full sun of early succession, ecology happens faster and at a more convenient height than in the woods. Sit for a few minutes in front of a clump of goldenrods in flower, and you are almost sure to see a nectar-slurping, multicolored fly come for a visit. If you are lucky and the visitor is not careful, a spider camouflaged to look like a flower might reveal itself in a most alarming fashion to have its breakfast. With a little bit more luck, or a lot more time, you might get to see a yellow-rumped warbler feast on the spider that captured the fly, and so on.

Although most old-field plants are insect pollinated, some still avoid having to feed pollinators by releasing their pollen into the air stream. All pollen grains are small, but those of wind-pollinated plants are really tiny. And, because the likelihood is so small that a pollen grain blowing in the breeze will land on a receptive female flower of the same species, wind pollinated plants each produce millions of pollen grains. Ragweed is such a plant, and it is justifiably loathed by hay-fever sufferers. At the height of the sneezing season, there can be as many as 1,000 ragweed pollen grains in a cubic yard of air. Wind-pollinated ragweeds often grow alongside insect-pollinated goldenrods, and many people blame the latter for the suffering caused by the former.

In addition to being irritating to the sinus cavities of the sensitive, pollen grains are slow to decompose, especially if they end up in the muck at the bottom of a lake. Because most plants produce pollen grains that can be identified by species, cores of muck from lake bottoms can reveal vegetation histories that span tens of thousands of years.

The pollen records from lakes in North Florida, for example, reveal that 18,000 years ago, when the continental ice sheet had pushed as far south as New Jersey, the vegetation was open oak woodland and the climate was cooler, drier, and windier. The annual layers of pollen also reveal that our Amerindian predecessors only started farming about 1,000 years ago; before that, they fed themselves by fishing, hunting, and gathering. Evidence of farming can be direct (such the presence of pollen from corn and other cultivated plants) or indirect (such as when the pollen of ragweed and other agriculture-following plants is abundant).

If the first people to colonize Florida arrived about 12,000 years ago and started farming only 1,000 years ago, then why do our lake-bottom mucks reveal a spike in the pollen of ragweed and other old-field species at about 15,000 years ago? It could have been some dramatic but temporary change in the climate, but it isn't obvious how plants – ones that now only appear after plowing – were favored by windy, rainy, or particularly cold spells. An alternative explanation that I find exciting is that the old-field plants of today, the ones we favor by farming, evolved in association with soil plowing proboscidians. Even if mastodons, gomphotheres, glyptodonts, and their ilk didn't dig for coontie tubers, passing herds must have churned a lot of soil, thus providing perfect seed beds for ragweed and other old-field species.

Until the extinct Pleistocene megafauna is reconstituted with modern genetic techniques and reintroduced, we'll continue to plow up the successional fields at the NATL on schedule. By plowing different patches yearly, biennially, and at other longer intervals, we will always have places on the UF campus where visitors can observe what much of our landscape looked like back when farming was widespread, or even farther back than that, when really large mammals roamed.

35. Apologies to a Native Fungus-Growing Ant

According to recent estimates, there are some 20,000 species of ants worldwide (family Formicidae, order, Hymenoptera) and about 220 species in Florida. There are harvester ants, slave-making ants, leaf-cutting ants, extrafloral-nectary-defending ants, and more variations on this six-legged theme than most people can imagine. When I was growing up during the Cold War, we recognized only two types: aggressive red ants and industrious black ants. I believe that our government approved pitting red ants against black, as long as the black ants won. Such is how we spent many a lazy summer afternoons when we grew bored with feeding ant lions and there were no older sisters to tease.

Since well before the Iron Curtain fell, Florida has been invaded by the South American fire ants (*Solenopsis invicta*), and the military-industrial-university complex responded with a truly frightening arsenal of ant-killing poisons. Way back in the 30s, calcium cyanide was the preferred fire ant poison. In the 50s the Sunshine State was sprayed with chlordane, dieldrin, and heptachlor. Starting in 1962, the year that *Silent Spring* was published, and continuing through 1978, 140 million acres were aerially dosed with mirex baits, but to no apparent avail. The open season on ants of all colors, personalities, and livelihoods continues to this day – just check out the arsenal of toxins available over the counter at your local "garden" store. This open warfare is unfortunate, given that most ants are not at all bothersome and many perform useful ecosystem services.

I write this chapter out of remorse for an act of formicide that I committed out of ignorance. My crime involved killing several colonies of perfectly friendly *Trachymyrmex septentrionalis*, having mistaken them for the dreaded fire ants. Workers of the two ant species are about the same size, and both make small mounds of soil around the entrances to their below-ground nests, but the similarities end there. Whereas *Trachymymex* consumes fungus that it cultivates on cut pieces of flowers and leaves that it carries down into its nests, fire ants are voracious predators of other insects, baby birds, wounded mice, lizards, snakes, salamanders, and slow-moving humans. Luckily for antdom, the mounds of *Trachymyrmex* colonies can be easily distinguished from those built by fire ants.

Knowing now that *Trachymyrmex* colonies typically build half-moon-shaped mounds, whereas fire ants' nests are more amorphous and usually have multiple entries, I will never again kill the wrong ants. I am also going to try to find and distinguish *Trachymyrmex* mounds at night by looking for the glow of fire flies (actually beetles, in the genus *Pleotomodes*) sitting at the openings of the ants' nests, where they spend their larval youths apparently glowing in the queen's otherwise-dark chamber (now isn't that delightful to consider?).

In the interest of self-preservation, and to protect my little feathered, slimed, scaled, and furred friends, I will continue to wage war on fire ants. If there was something I could do on behalf of the *Trachymyrmex*, I would certainly try. Perhaps increasing awareness of these interesting little beasts' habits goes some ways towards making amends for my unwarranted formicide.

A potentially exploitable chink in the armor of fire ants is revealed when their habitat floods, presumably a reflection of their evolution on the floodplains of the Amazon and other South American rivers. When a fire ant nest is submerged, the beasts emerge, grab onto one another, and make baseball-sized ant masses with the queen or queens in the middle. These masses of ants float until they lodge on solid ground as the floodwaters recede. When our lower field floods, as it has several times over the past thirty years, fire ant colonies bob around on the surface in a most enticing manner. In that state, they would seem imminently susceptible to control efforts, but all of mine have failed.

A ball of fire ants can be picked up on a canoe paddle and tossed, which is amusing but not very effective as a control measure. Burning the balls with a propane torch is also frustratingly ineffective – while the outer layer of ants is easily toasted away, thousands more remain to protect the queen and the colony. Exploding them with firecrackers is exciting but painful – it's hard to paddle far enough away to avoid the flying and very annoyed beasts. Dropping the ant balls in soapy water also doesn't work very well. Obviously, I have yet to devise an effective approach.

Despite my efforts, after each flood there's an evident line of new fire ant nests at the high water mark. In anticipation of the next flood, I will appreciate suggestions, especially those proven to be successful and at least moderately safe.

36. Ardisia and Empty Niches

Ecologists often conceive the natural world as being divided into niches, with one niche per species and one species per niche. A niche can be thought of as the combined address and occupation of a species: that is, where it lives and what it does for a living. For example, big-leaved pawpaw is a forest-dwelling species that makes due with the sparse light filtering through the overtopping canopy. In contrast, narrow-leaved pawpaw is a light-demanding species of open habitats. In a world that is being rapidly homogenized by humans moving species from their ancestral lands to new places on the planet, the niche concept might be useful in explaining why some introduced species successfully invade their new homes.

People who are alarmed by the unprecedented rate of species extinction often hate introduced species with a purple passion. Indeed, invasive plants and animals, including cats and cows, rank second behind habitat loss as the most important cause of species extinction. Seeing what kudzu, cogongrass, hydrilla, melaleuca, Chinese tallow, or water hyacinth can do here in Florida, it's easy to believe that exotic invasive species are a global menace. That they can invade disturbed habitats, such as roadsides and mine spoil heaps, is easy to understand, but it's harder to explain how they can invade what appear to be intact ecosystems.

When an introduced species successfully colonizes, one often-invoked explanation is the empty niche hypothesis. Although the idea of a niche being empty inspires me with melancholy, it might explain why Japanese climbing ferns and air potatoes have so effectively invaded our forests and why cogongrass is such a threat to our savannas: it's because these areas respectively lacked ferns that climb, vines that produce aerial tubers, and fire-favoring grasses that spread vegetatively. I sometimes wonder whether empty niches are only revealed after they are filled, but the hypothesis might yet have some explanatory power.

Soon after I became aware of the substantial environmental impacts of escaped populations of ardisia, I was at a barbecue at the home of the famous ecologist, the late H.T. Odum. I was walking with H.T. and a gaggle of students along the trail bordering the sinkhole behind his house when I saw an ardisia with ripe red fruits. Without hesitating, I reached down to extract it from the otherwise-natural setting. H.T., who was a looming figure in many ways, reached down just as quickly to restrain me. I stood agape as he defended what I considered a vile intruder. He viewed ardisia as a contributor to ecosystem functioning and a filler of an otherwise vacant niche. In no uncertain terms, he told me to leave it be.

Ardisia crenata, known variously as Christmas berry, coral ardisia, and just ardisia, is a small shrub or treelet native to Asia that was introduced to Florida in the 1920s for horticultural purposes. In many ways, ardisia is the perfect landscaping plant. It seldom grows more than about three feet tall, its dark evergreen foliage suffers little from herbivores or pathogens, its white flowers are lovely, and its large red fruits are held in attractive displays that persist for months. The species is tolerant of deep shade and easily propagated by seed. Once established, it requires little irrigation and no fertilization. Finally, dense patches of ardisia seem impenetrable by taller-growing trees or vines. Despite its favorable features as a yard plant, I view ardisia as an insidious invader of otherwise-intact hardwood hammocks.

Ardisia is not your typical exotic invasive species. Its seeds are large, not small like many ruderal species. It is also very shade-tolerant, which is not characteristic of most weeds. Finally, it grows slowly and lives long. Where ardisia comes to dominate an area, it does so by working its way slowly through the understory until all that's left is ardisia.

Although where ardisia grows in abundance nothing else does, I am not aware of it driving any native understory species extinct. I must also agree with H.T., at least insofar as ardisia does cycle nutrients, transpires water, fixes carbon, and otherwise participates in ecosystem functions. The reason that I still can't abide by ardisia in natural areas is this: knowing humans introduced it spoils my experience of nature. If I want to see species from other parts of the world, I can travel abroad, or at least visit a botanical garden. But when I am in Florida, I don't want to see beer cans, I don't want to see oil spills, and I don't want to see biological pollutants like ardisia.

My concern about the homogenization of the world is not restricted to worries about the increasing rates at which *Homo sapiens* benefit a few species at the expense of many. I also worry about the comparable rates at which languages and cultures are being lost. I don't frequent the McDonalds in Singapore, and I don't want to see pines in South Africa or eucalyptus in California. But I'm far from being a purist: our garden hosts onions from Asia, sugar cane from New Guinea, and potatoes from the Andes. It's just that I feel a loss when local diversity is overwhelmed by introduced species, even if vacant niches are filled in the process.

37. Would You Prefer Red Coontie or White?

While I admit to having only vague ideas about the restaurant fares of yesteryear, if I were dining out in Florida a few centuries back and the server offered me red coontie or white, I'd opt for the red. There's admittedly some exoticism associated with white coontie, derived as it is from the stem tubers of a very primitive plant, but I worry about the long-term health impacts of ingesting it in quantities. Then again, based on my admittedly unresolved experimental archaeology research, red coontie appears to be one of those "starvation" foods: the more you eat, the more malnourished you become. But then what exactly was the coontie that was the staple food in Florida before it was dropped from the menu by the colonists from across the Atlantic, and how was it prepared?

I'm hesitant to eat white coontie made from our native cycad, *Zamia floridana*, mostly because it contains cycasin alkaloids (which cause liver damage) and beta Methylamino-L-alanine (the cumulative effects of which are senile, dementia-like neurological symptoms). Until the life expectancies of cycad-consuming Australian aborigines and Pacific islanders increased, this form of psychosis was mistaken for alcoholism. I'm aware that, by grinding up and then leaching the stem tubers, at least some of the toxins can be removed, but I worry about consuming even trace amounts. For me, white coontie starch is fine for stiffening collars (for which it was once used commercially), but it is not high on my list of favorite foods. Although, from the plant's perspective, these toxins provide protection against herbivores, there are specialized species that eat little else, including the larvae of the glorious atala hairstreak butterfly.

What some people don't realize is that, when it comes to coontie, there are choices. The word coontie itself isn't very revealing insofar as it just means "flour root" in Creek. This confusion is also understandable given that *Zamia*, the cycad, is generally referred to as Florida coontie. Less well known is red coontie, which is made from the swollen, belowground stems (i.e., rhizomes) of one or several of the dozen species of catbrier (*Smilax* spp.) native to Florida.

Due to the thorny stems of these pesky vines, *Smilax* species are generally referred to as catbrier or greenbrier, but some stout-stemmed species are called bullbrier or bamboo vine. Several South American and Caribbean species of *Smilax* were exported to Europe in astounding quantities from the 17th to 19th Centuries, mostly as an herbal treatment for syphilis. The catbrier of former pharmaceutical importance is known by its Spanish moniker, *sarsaparilla* (derived from the words *sarsa* for prickles and *parilla* for climber). *Sarsaparilla* was, and perhaps still is, also used as a foaming agent in soft drinks. So far, I haven't been able to find out much about how *sarsaparilla* was or is commercially grown and processed, but I am interested.

Whereas the writings of scientists often reveal profound knowledge on a variety of topics, on the subject of coontie I have more outstanding questions than brilliant insights. Although laboratory analysis revealed that catbrier tubers contain as much starch as "Irish" potatoes (which originated in the Andes), I have still not perfected the technique for rendering the rhizomes into food. I shouldn't shirk responsibility for my failures, but progress on this project has been impeded by the poor attitudes of two of my would-be collaborators. Although my son is often a willing helper and subject, neither his older sister nor his mother have been at all cooperative. For example, I expected my wife and daughter to be thrilled with the live oak mortar and pestle that I so painstakingly and lovingly made for them, but my vision of the two ladies happily pounding red coontie and perhaps singing while I reclined in my hammock with a beer was merely an illusion; they would neither pound nor even sample the bright red coontie pancakes Antonio and I so painstakingly prepared and offered. I admit to not yet perfecting a method for removing the tannins and other astringent compounds that make red coontie red, but they could at least have taken a taste. After all, I followed the explicit instructions of the famed naturalist William Bartram, about whom I now harbor some suspicions.

In his magnificent *Travels*, Bartram, known among the Timucuans as *Puc Puggy* (the "Gentle Flower"), described in substantial detail how to make bread from catbrier rhizomes:

> *"They chop the roots in pieces, which are afterwards well pounded in a wooden mortar, then being mixed with clean water, in a tray or trough, they strain it through baskets; the sediment, which settles to the bottom of the second vessel, is afterwards dried in the open air, and is then a very fine reddish flower or meal: a very small quantity of this mixed with warm water and sweetened with honey, when cook becomes a beautiful, delicious jelly, very nourishing and wholesome. They also mix it with fine corn flower, which being fried in fresh bear's oil makes very good hot cakes or fritters."*

After trying his recipe, and many variations on the theme (without the bear oil of course), I can't help but wonder whether the braves who described the process really knew of what they spoke. Mightn't they have left out a step due to their own ignorance or to protect a trade secret?

Having heard that coontie preparation sometimes involved fermentation, I brewed up a batch of red coontie beer. Julia Morton, that pillar of economic botany, mentioned in one of her books that early settlers mixed red coontie rhizomes with molasses and parched corn to make root beer. Unfortunately, the results of my efforts were less than salutary; even I admitted that the brew was awful, worse even than my muscadine wine. One problem with my catbrier brew might have been that it fermented very slowly, which allowed plenty of opportunities for colonization by microbes other than beer yeast. The process could have been accelerated by adding honey or sugar, but since honeybees, sugar cane, and sugar beets were all introduced to the Americas by Europeans, I opted for historical purity. And I ended up with biochemical diversity. I suspect that my red coontie beer contained some of the ethanol for which I hoped, but with liberal admixtures of methanol, aldehydes, ketones, and other toxins.

I drew inspiration for my next approach to red coontie from my African ancestors who started playing with fire about 1.9 million years ago. Lots of experts doubt that australopithecines, the ape-like ancestors of *Homo erectus*, had the brain-power to control fire and, if they did, used it only to roast meat. On the other hand, some human paleontologists credit fire use for the rapid evolution that transformed apes into recognizable humans. Not only that, they think that, in addition to meat roasting, fire was also used to render roots, tubers, corms, and rhizomes into edible food. Unfortunately, when I chopped and then cooked the rhizomes – making sure to change the water frequently – I was no more successful than I was at making coontie beer.

While on sabbatical in the south of France, I was confronted with the same dual problem of bitterness and fibrosity when I tried to make freshly fallen olive fruits into something edible. From the web, I'd extracted several recipes for treating those astringent and rock-hard little fruits. Most of the recipes called for repeated leaching in strong brine solutions for several weeks, but an attractive shortcut was to soak the fruits in a lye solution for a scant three hours. With dreams of tapenade, I tested fourteen different recipes. The principal result of all of them was newfound respect for olive makers. Of the twelve batches I ran with different salt solutions and flushing frequencies, the resulting olives were too soft, too hard, or too astringent. The two jars of fruit that I soaked in lye for different lengths of time, in contrast, had great consistencies and no astringency, but they always tasted like soap, even after repeated flushing with fresh water. Although I failed at olive making, the experience got me thinking about how lye might help render other wild foods edible.

Lye strong enough to peel paint is made simply by leaching wood ash in water. Unless you want to remove your fingerprints, I recommend that you do not stir the mixture with your bare hands. Certainly Amerindians had ready access to lye, but did they use it in their preparation of coontie? My initial experiments with lye treatments have been encouraging. Pounded *Smilax* tubers soaked in lye do indeed soften and surrender their tannins rapidly. What eludes me, as it did with French olives, is how to remove the residual lye taste from the rendered mass.

Despite having not yet succeeded in my culinary experiments, I think I know which of our catbrier species was the principal source of red coontie for both the local Timucuans and the Seminoles who followed. The scientific name for this thick-stemmed and nearly thornless species is *Smilax smallii*, named after John Kunkel Small, author of the first comprehensive flora of the Southeast. In the vernacular, this species is sometimes known as jackson vine and lanceleaf in addition to its more-generic name, greenbrier. The foliage of jackson vine is evergreen and sufficiently attractive that it was commercially marketed for use in Christmas decorations through the 1960s. Large quantities were shipped northwards by rail from Jacksonville every year during the Yuletide season, a trade that has now apparently stopped.

I can't yet distinguish the taste of red coontie derived from different species of *Smilax*, but I do know that, pound-for-pound, *S. smallii* is the easiest to excavate. Using just a stout stick with a fire-hardened point, I excavated nearly 100 pounds of tubers in a sweat-drenched half-hour. In contrast, harvesting a similar amount of the next most likely candidate, *Smilax laurifolia*, takes about four times longer, even with a long-handled, metal-bladed spade. The least likely candidate is perhaps our most common: the starch-containing rhizomatic swellings of *S. bona-nox* are small, widely dispersed, and seemingly impossible to harvest in meal-sized quantities. I should also point out that jackson vine is common near sinkholes, just the sorts of places Amerindians once frequented.

I wonder whether the prodigious size of jackson vine rhizomes is partially the result of thousands of years of selection by native Floridians. Although they may not have started hoe wielding and row cropping until only 1,000 years ago, they almost certainly promoted the growth of their principal food plants. For red coontie, it would be easy to increase the supply by sparing some portions of harvested rhizomes and transplanting them to convenient places. The dozen catbrier rhizome fragments that I transplanted over the years have all flourished, sometimes to my dismay.

Whichever *Smilax* species native Floridians ate, I expect that they selected for genotypes with favorable characteristics, such as low tannin contents. Or perhaps they promoted the growth of lots of different species and varieties and, when one would dine out in Timucuan times, the coontie steward might have offered a range of vintages, each with its own unique taste and bouquet.

38. Root-Sucking Mosquitoes

Walking with the now-late Merle Kuns through the oak hammock at a retirement community of the same name, we stopped to inspect a weird contraption beside the trail. I was scoping out the area as a potential field site for my undergraduate ecology class, and Merle was my guide. A self-described "inmate" at Oak Hammock, Merle was a retired medical entomologist, P-51 Mustang pilot during World War II, veteran of the campaign against hemorrhagic fever in lowland Bolivia back in the 60s, naturalist, photographer, and all-around interesting fellow. I was correct in my surmise that he would have a tale to tell about the contraption, but I didn't realize that my simple question of "What's that?" would catapult me into a spate of learning about the lives of mosquitoes.

In reference to my query, Merle explained that the machine in question was a "Mosquito Magnet" contributed by Gainesville's Mosquito Control Program. These increasingly common devices burn propane to emit a combination of carbon dioxide and heat that, together, attract female mosquitoes into a death trap. He mentioned that the particular mosquito that had so plagued his community after several recent hurricanes was *Mansonia dyari.* He then added that these beasts had invaded Oak Hammock from Bivens Arm, a few miles to the south as a mosquito flies.

I was impressed that he knew the species of mosquito (only vaguely aware that mosquitoes flew that far), though I couldn't figure out how he knew the breeding site of this particular mosquito. But I first asked whether the device worked. Merle responded that the trap did indeed catch a lot of mosquitoes, but when I pushed him about whether the effect was sufficient to allow pest-free walking, he was less committal.

We walked on for a ways, me cogitating about the rest of what Merle had said and him patiently awaiting my question. I finally asked how he knew that the *Mansonia dyari* plague was not of more local origin. He explained that, while there are dozens of mosquito species that breed in nearby sinkholes, tree cavities, discarded tires, and other water-holding spots, *Mansonia dyari* larvae are obligate associates with the roots of water lettuce (*Pistia stratiotes*), and that the nearest population of this floating aquatic plant was Bivens Arm.

Given how much I deal with mosquitoes, both here in Florida and while conducting my research in the tropics, it's embarrassing how little I know about their natural history. For example, I should be able to recognize the carriers of West Nile and St. Louis encephalitis, diseases of some local concern, as well as the vectors for dengue and malaria. Malaria has been eradicated from Florida, but my father caught a bad case in 1942 at the Army Air Corps training facility in Bushnell just before shipping off to India. I still remember him having tremors in our kitchen in the 1950s, but his first attack came as he walked down the gangplank in Calcutta. Doctors there thought they had a new record for gestation of the malarial plasmodium, but then realized that his case was Floridian. To help me out of the mire of my ignorance, Merle recommended that I start my education about these infernal beasts by reading an old book by a friend of his entitled *The Natural History of Mosquitoes*.

Merle's friend was the famous naturalist (and 1934 UF graduate) Marston Bates. His 1949 book on mosquitoes is a gem that he'd written in the "small Colombian town of Villavicencio, east of the Andes and west of nowhere." I learned from Bates that the swimming larvae of most mosquitoes spend their time hanging upside down, with their breathing tubes (or siphons) piercing the surface skin of water. Only when threatened do they dive to the bottom with awkward-looking twitches of their bodies, quickly rising back to the surface to get a breath when they need it. Not so with *Mansonia dyari*, the larvae of which insert their toothed siphons into the submerged roots of water lettuce. Apparently, sufficient oxygen diffuses down into the soft, spongy roots to keep both the root tissues and mosquito larvae alive. I can imagine that the larvae benefit from this association because it allows them to escape surface-searching predators, but I have to wonder about the history of this species-specific association.

Whether or not water lettuce is native to Florida is still debated. Although William Bartram described mats of it on the St. Johns River in 1765, it might very well have been introduced earlier in the ballast of sailing ships. The Florida Exotic Pest Plant Council has this species on its Category 1 list of plants that invade and disrupt natural ecosystems, and it is generally considered one of the world's worst weeds. Seems to me that, if water lettuce was introduced, so was *Mansonia dyari*, and thus we are dealing with two exotics, one that clogs our waterways and the other that sucks our blood.

39. Cogongrass

When a friend with road grader offered to smooth the teeth-rattling washboard road that leads into Flamingo Hammock, we jumped at the chance, unaware that he doesn't practice safe bulldozing. For several months after he graded, the ride was mercifully smooth. But then, two years later, when the potholes were again starting to deserve individual names, the bright green blades of cogongrass became evident along the roadside. Apparently, he'd unknowingly propagated this dreaded weed when rhizome fragments were carried from another job on the tracks of his machine.

Cogongrass, known to science as *Imperata cylindrica,* is an Asian species that is spreading like wildfire over much of the world, including Florida. It was purposefully introduced here as a forage crop, but cattle will eat only the young sprouts – which emerge soon after fires – because its mature leaves are impregnated with sharp, gut-cutting and teeth-wearing crystals of silica. Too bad about its inedibility, because, although it prefers sun and likes lots of nutrients, cogongrass grows well under a wide range of conditions, including the poorest soils and mine tailings; in fact, it isn't slowed down by much of anything except the deepest shade.

If cogongrass would just fit in amongst our diversity of native fire-favoring herbaceous species, it wouldn't be such a problem. Unfortunately, cogongrass grows faster, taller, and in denser swards than the native grasses and forbs that it crowds out. When it burns, which it does readily, it burns hot, high, and thoroughly. When a cogongrass sward catches fire, few patches remain unburned, and even head-high longleaf pines caught bolting from their grass stage are toasted.

Even when it isn't burning, cogongrass is an aggressive species. Its roots are efficient at scrounging nutrients, which makes the species an effective belowground competitor in the poor soils that abound in our area. Its rhizomes, which look like fat roots but are really belowground stems, grow in incredibly dense tangles and are chock full of sugar, which allows the plant to resprout repeatedly after being cut or burned. Worse yet, the tips of these rhizomes are sharp and can pierce the roots of other species. And aboveground cogongrass is no slouch: it readily shades out virtually all of the hundreds of flowering plants native to our pine savannas.

Given its ability to invade intact pine savannas, to outcompete pasture grasses, and to otherwise wreak havoc in both natural and anthropogenic ecosystems, it's no wonder that the Florida Exotic Pest Plant Council lists cogongrass as a Category I Invasive Species, while the USDA considers it a Noxious Weed, and the Plant Conservation Alliance lists it as an Alien Invader. It's also ranked as the 7th most noxious weed worldwide by people who rank that sort of thing. With these accolades, it's no surprise that a great deal of research effort has been expended on ways to kill cogongrass.

Small infestations of cogongrass can be controlled by pulling the roots up every time a leaf appears. With due diligence, you might be cogongrass-free after several years. If your infestation is larger or your time and energy are more limited, there are several chemical solutions to the cogongrass problem. Of course, these herbicides need to be applied several times to kill the rhizomes. And it goes without saying that these chemicals are frightfully expensive. It should probably also be pointed out that, before your cogongrass is dead, you will have also killed every other plant in sight, which means that you'd better be ready to quickly replant something to protect your site from other invaders.

I first encountered cogongrass as a Peace Corps volunteer in Malaysia, where it's actually native and known as *alang-alang*. Among the government foresters with whom I worked, *alang-alang* was despised as an impediment to industrial tree plantation establishment in what they considered wastelands created by subsistence farmers who slashed and burned without apparent regard for nature. To promote rural development and to reforest the land so as to make it once again productive, they used World Bank and other development-agency funds to hire those same farmers to plant exotic acacias or pines in lines through the *alang-alang*. Of course when the designated areas were fully planted, large work forces were no longer needed and jobs were cut. Miraculously, soon after this happened, the young plantations would catch fire and have to be replanted, often with funds from a different donor bent on reforestation and social welfare enhancement. Meanwhile, the local people with whom I socialized mostly liked *alang-alang*; they used it for roof thatching, they cleared their agricultural fields in it, and they otherwise didn't view their landscapes as wastelands.

I relate this Asian anecdote as a reminder that all species are native somewhere. The other message is that a "weed" is just a plant growing where some human doesn't want it to grow; even cogongrass has its admirers. And finally, while I don't want to diminish the very real threats posed by cogongrass in Florida, we share cultural biases for trees and forests and against grasses and savannas. But if you do see an erect grass with half-inch wide blades and a prominent midrib that is whitish and a bit off-center, please do something before it takes over.

40. Crown Shyness

Out of general friendliness but with an eye to academic bridge-building, my geneticist colleague Rob Ferl invited me over to his house for a barbecue a few decades back. Back then, Rob was just another junior professor in my department, but now he's famous. He seems to run the entire biotechnology program at the University of Florida. Of the twenty-or-so people present at this barbecue, I suspect I was the only one who wasn't a molecular geneticist or biochemist. As an organismic biologist – someone who mostly thinks about the wholes rather than the parts of plants and animals – I should have felt a bit out of place, if not outright threatened. After all, these lab types are often assumed to be the sworn enemies of whole-organism researchers, competing as we do for funding, faculty lines, space, and recognition. Instead, with a beer balanced on my belly, I comfortably reclined on a hammock strung between two slash pines while the molecular types discussed the genes and enzymes on which they worked.

Before fortune, or at least fame, caught up with Rob, he lived in a housing development in a former slash pine plantation. Thorough ditching and draining, which remains permissible because flatwoods were loopholed out of legal protection as wetlands, the area only floods during real frog chokers of storms. On the house lots in Rob's neighborhood, many of the twenty-five or thirty-year-old row-planted pines were left standing. I'm not sure who decided to retain the closed stand, but they showed some wisdom.

Like most trees, slash pines grown in crowded strands become tall and thin. Radically opening such stands with what (in the trade) is called a "suburban thinning" subjects the remaining trees to wind-induced mechanical stress beyond what they can withstand. In contrast, Rob's trees seemed to be doing pretty well, swaying together in a light breeze that penetrated down to my hammock.

After the guests had heard enough about each other's enzyme systems, a geneticist politely asked about mine. The question was initially a puzzler, but then I responded that I work on ecosystems not enzyme systems. I thought that my answer was perfectly clear, but a biochemist asked for a bit more detail.

It was that time in the evening when you stop looking up at the trees themselves and start looking at the spaces between them, so I pointed up towards the tree crowns and explained that I studied the mechanisms creating the spaces between trees. It's not the major focus of my research, but I had just completed a study of what are called crown shyness gaps. In the study, conducted in a Costa Rican mangrove forest, a couple of students and I found that these openings on the borders of tree crowns are created when the trees bang against each other in the breeze. The breakage and abrasion of peripheral branches keeps the tree crowns shy of one another, or at least prevents them from interdigitating. In closed canopy forests, crown shyness gaps are important to understory plants and to canopy-dwelling animals that can't fly. Understory plants are typically starved for light, and the sunflecks that pass through crown shyness gaps and other holes in the canopy are critical to their survival. For non-volant canopy animals, moving from tree to tree requires crossing these shyness gaps, which can be several feet wide in windy forests.

For reasons that elude me, the lab biologists around me seemed embarrassed by my explanation of the nature of my work. Admittedly, crown shyness isn't the main focus of my research, but I thought they would find the phenomenon fascinating. Instead, what followed was one of the best examples of a pregnant pause that I have ever witnessed. Luckily, Rob had been following the rock-bound drift of our conversation and was quick to respond to the situation by calling for dinner, which actually still wasn't quite ready.

41. Living with Vines

The most pernicious weeds in our vegetable garden are vines. Despite various attempts at eradication, trumpet creepers, yellow Jessamine, wisteria, and grapevines still sprout up between the collards or tomatoes, depending on the season. The fact that these vines are a legacy of a failed experiment that I conducted on the same site many years ago only increases my frustration with their resilience.

I am primarily employed as a scientist – a plant ecologist, to be more explicit. One of my specialties is vines, like those plaguing our gardening efforts. Although I doubt that my children brag about it at school, I may know more about vines in general and vine roots in particular than anyone else in the world. I study them and write technical papers about them, but I have not yet learned how to live with them.

Vines rise in the world by relying on other plants for mechanical support: in other words, they are structural parasites. Freed from the need to be self-supporting, vine stems can remain thin but grow in length at extraordinary rates and to extraordinary extents. Because of their exceedingly efficient plumbing, vine stems can supply as many leaves with water and nutrients as tree stems many times larger in diameter. The records folks at Guinness haven't yet picked it up, but I measured a vine in Panama that was nearly two feet thick and grew in the crowns of sixty-four canopy trees, descending to the ground and growing back into the canopy twelve times in the 1.2 acres of forest that it covered. With their looping and dangling stems, vines are lovely to look at, and they would be a nice addition to almost any landscape if they would only stay where we put them and otherwise play nicely with the other plants.

Along with me, trees and power-line managers probably hold vines in equal disdain. With their phenomenal capacities for stem elongation and leaf production, vines can quickly overwhelm and smother trees and power lines alike. The folks at our local electrical utility tell me that many of our power outages are due to vines, mostly grapes and air potatoes making electrical connections between lines. Similarly, where vines are abundant, trees are bent and grow slowly. Trees freed of their vine burdens typically grow twice as fast and produce up to an order of magnitude more fruit as their encumbered neighbors. And I wish you good luck if you try to kill the vines causing an infestation. Unless you resort to herbicides and get the formula correct, you're in for a long and drawn-out battle. Not only do vines readily sprout from cut stumps, stems that fall from the canopy also often take root, sprout abundantly, and climb back up wherever trellis supports are available.

The study that resulted in the persistent resprouting of vines in our garden was designed to address the age-old question of whether vine roots are more efficient foragers for nutrients than tree roots. Unless you are used to thinking in plant time, the idea of roots foraging might seem like a stretch. But as pulses of nutrients come and go in space and time, the spoils go to the swiftest, even if "swift" in this case refers to growth rates of an inch or two per day. The logic behind this experiment was that, just as vine stems can be thin because they rely on other plants for mechanical support, they should also not need thick roots for mechanical purposes. If this is true, then, for a given investment belowground, the thin roots comprising a vine root system should explore soil more efficiently and more rapidly than do the roots of trees. In any event, that was the logic behind the rather elegant experiment I conducted, with the able assistance of my mother.

The race between vine and tree roots to nutrient baits involved four tree species (magnolia, laurel oak, Carolina cherry, and loblolly pine) and the four vine species that are still cropping up amongst my crops. We planted seedlings in a grid, with vines and trees equidistant from perforated PVC tubes sunken vertically in the ground and filled with sand and slow-release fertilizer. To keep out rain, we constructed a plastic roof over the entire experimental plot. Every day, it was Mom's job (she was retired and living with me at the time) to pour water into the top of each of the sixty-four tubes. Theoretically, water moving down the tube would pick up nutrients and carry them out into the soil, thereby creating a nutrient-concentration gradient up, which would entice the vine and tree roots to grow. By pulling up and inspecting the tubes frequently, I could (again theoretically) determine whether tree roots or vine roots were the first to colonize the nutrient bait. Unfortunately, for the experiment and for maternal relations, the vine roots grew exceedingly rapidly but did not restrict themselves to growing up the nearest nutrient concentration gradient, perhaps because the expected nutrient concentration gradients did not develop. And then there was the problem with the roof of plastic sheeting; how was I to know that we should have specified ultraviolet light-stabilized plastic? After a few weeks, vine roots were everywhere, the roof was disintegrating, and Mom was disgusted with her Ph.D.-holding disaster of a son.

A few years after this fiasco, with Mom happily residing in Flagler Beach, I redesigned the experiment so as to avoid the garden site pitfalls. In this more controlled study, I constructed long, narrow plywood boxes, one per plant. In each box I created the desired nutrient gradient by installing temporary partitions and backfilling with sand mixed with different amounts of fertilizer. The nutrient-rich end of each "root-race" box was enclosed with plexiglass that was kept covered except while checking for roots. Just as I expected, vine roots outgrew the tree roots in nearly every case. These results help explain why vines are as good competitors with trees belowground as they obviously are aboveground.

I've been sitting on these root-race data now for about a decade, and really should publish them. Although I doubt that the publication will win me a free trip to Stockholm, I also can't imagine how many people will be relieved to learn that vine roots really do forage more efficiently than tree roots. Unfortunately, when I finally went to retrieve the data, they were on a five-and-a-half inch floppy disk, coded in Lotus 1-2-3, and described in a Word Perfect document. Fortunately, my colleague, Jack Ewel, an emeritus professor of ecology, came through with a disk reader and the required program translations. Now the only thing standing between me and fame among the root foraging experts of the world (all three of them) is sloth.

42. Armadillo Croquet

I can't help but wonder how Hutch's brain works. Imagine armadillo croquet. The rules he developed are simple. When an armadillo is detected, usually by the sounds of their loud snorfling in the leaf litter, all contestants slowly surround it in a circle. Armadillos are fairly oblivious, so close approaches are quite possible. Eventually, though, even armadillos spook, jump, and then run off in what seem to be random directions. To win a round of Armadillo Croquet, the beast must pass between your spread legs. Any number of contestants can play and, because no harm is done, the game should be approved by the Society for Prevention of Cruelty to Animals.

Archie Carr, renowned scientist and nature writer, would not have approved of armadillo croquet. For him, the only good nine-banded armadillo in Florida is a dead one. He blamed these introduced, nearly toothless but voracious carnivores for the decimation of our beloved native herpetofauna (i.e., amphibians and reptiles), the declining populations of ground-nesting birds, and many other environmental and social ills. I became aware of his loathing the first time I accompanied him on a field trip.

Archie was driving, I remember that clearly. The armadillo was in the other lane. Its demise was swift after Archie crossed the double line. I didn't die. Disregarding his ashen-faced passenger, Archie pulled off the road, jumped from the cab, and fetched his prey. Upon returning with the carcass, and in the way of an explanation for his seemingly erratic behavior, he said it was to feed his bald eagle. Later, I learned that he had an eagle feeder behind his house in Micanopy. Wild eagles knew to check at the Carrs for tasty treats, and they often arrived before turkey vultures and black vultures, their main competitors for carrion.

Armadillos are peculiar beasts, dead or alive. The first Spaniard to describe one to Queen Isabel undoubtedly rued the day he first saw the creature with a "snout of a pig, shell of a turtle, and head of a squirrel." I say this with some confidence, given that he was apparently thrown in the pokey for mocking his monarch with his tale of this strange New World beast.

To the description provided to Isabel, the fact that there are twenty extant species of armadillos, all native to the Americas, might be added. They range in size from much smaller than a breadbox to a bit smaller than a mid-life crisis sports car (and are also pinkish). All armadillos have strong claws and grinding molars, but no canines or incisors, so they are not biters. They eat mostly insects, but are also fond of eggs, worms, and salamanders. Powerful diggers but lazy, they often round out the holes excavated by gopher tortoises to make their burrows. Cold winters are hard on these nearly-naked mammals, both because they freeze easily and because their prey become scarce, at least near the soil surface.

Archie's claim that armadillos tip the balance of nature is based on the fact that they were recently introduced to Florida, several times in fact. Two are known to have escaped from a private zoo in South Florida and another pair was released by a Marine from Texas after World War I. But even without the direct help of humans, armadillos were bound to recolonize Florida. I say "recolonize" because, as little as 10,000 years ago, Florida was home to several now-extinct species, one of which (*Holmesina septentrionalis*) reached six feet in length and weighed more than 500 pounds. Biogeographical expansion by nine-banded armadillos was also facilitated by their ability to cross water bodies, either by walking across the bottom or by floating on the surface after filling their stomachs and intestines with air.

Populations of nine-banded armadillos fluctuate radically from year to year. Given that they give birth to identical quadruplets, population explosions are no surprise, but what keeps their populations in check is another matter. Cold winters are part of the story. Armadillos also succumb to leprosy, but often only after carrying it for up to four years. Another obvious cause of dilly deaths is automobile encounters. If they hunkered down when frightened, most vehicles would pass them over, leaving them unscathed. Instead, even with natural selection working overtime on the behavior, they jump up to three feet straight up into the air when startled, making Buick bumpers an obvious cause of armadillo mortality. For keeping the world safe from the "little armored ones," we can also thank coyotes, foxes, dogs, panthers, alligators, and humans with refined palates.

Soon after purchasing what became the core of Flamingo Hammock, we were both plagued by an armadillo outbreak and burdened by massive mortgages. I can't remember if it was Hutch or Richard who came up with a solution to both of our problems. In any event, the solution involved Richard's smoker and Hutch's converted school bus. The smoker was fashioned from the all-metal Kelvinator that had graced Richard's mom's kitchen for many years.

Our plan was to load the smoker into the back of Hutch's old school bus, pile in a few face cords of dillies, and then deliver – to fancy restaurants in Atlanta and D.C. – armadillos we'd smoked along the way. We figured that rich folks in those cities would pay pretty pennies for armadillo on the half-shell, even if it does just taste like chicken.

I am sorry to report that this was another inspired plan that went awry, due mostly to lack of discipline. Instead of making a profit with the Kelvinator full of armadillos cool-smoked to perfection, we ate our capital. The shells cracked and the wine flowed, and I remember bits of a heated theological discussion about armadillo feet and whether their consumption could be Kosher. I do not recall how that debate was resolved, but the next day, my head hurt and our budding business was in shambles.

43. Suburban Acts of Predation

Waiting at a traffic light at a busy intersection in front of a Publix supermarket, my attention was drawn by a flurry of feathers and fur. A red-shouldered hawk (*Buteo lineatus*) swooped down from I-don't-know-where and nabbed a grey squirrel (a.k.a., a brush-tailed tree rat) right there on the sidewalk. The squirrel seemed too big for the hawk, but before the light turned green, the bird was aloft with its prey.

When I was a kid back in the DDT-laced 1950s, when the springs were getting silenced, red-shouldered hawks were rare. Like other species at the top of the food chain, high tissue titers of pesticide precluded their laying eggs with shells thick enough to bear the weight of the brooder. Habitat destruction was perhaps an even a bigger reason for the 60-98 percent decline in red-shouldered hawk populations since the European colonization of North America. The species, in the literature at least, is reported to require large tracks of mature-to-old-growth forest. What's going on, then, with all these suburban red-shouldered hawks? Is it the super-abundance of grey squirrels that is fueling the population increase? Have they evolved to tolerate our presence, even our exhaust-spewing automobiles? I hope red shouldered hawks are indeed on the rebound, and I admit to feeling little remorse for the tree rats that are fueling their comeback.

Over the years, I have been collecting data on public reactions to suburban acts of predation. In particular, whenever I come across a group of students watching a hawk stalk or bag one of the brush-tailed tree rats that abound on the UF campus, I ask as many of the gawkers as I can whether they are rooting for the predator or prey. To my surprise, the data reveals a strong female gender bias towards the predator. Rather than chance offending anyone, I won't explore what this might mean about the hearts and minds of UF's co-eds, but it does make you wonder. I will admit that I too am a hawk supporter, perhaps because brush-tailed tree rats decimated our bean crop once again this spring.

The suburban predation event that I have not yet witnessed is a coyote eating a feral cat. Although the howls of coyotes are now regularly heard throughout the sandy lands of the South, the beasts themselves are seldom seen. What they are eating is still something of a mystery, but small pets apparently figure prominently in their diets. A wildlife biologist I know is convinced that the best way to monitor coyote populations is by counting lost-kitty posters stapled on telephone poles. We have a pack that passes near our house at least once per fortnight, and feral cat populations are now scarce. Perhaps it is just my imagination, but the dawn choruses of songbirds seem to have grown more intense of late. And my wife complains more often about the incessant calls of the chuck-wills-widows through which I peacefully sleep. Even if it is just my imagination, there's no reason to doubt that, by eating feral cats, coyotes are good for our feathered friends, Roadrunner notwithstanding.

But the largest predator that roams our suburbs is the American alligator, an exciting beast, to be sure. While I bemoan the losses of human lives and limbs to these almost cold-blooded beasts, I can't help but be thrilled that there are still some places in the world where reptiles rule. Alligators and their kin actually did rule the world for more than fifty times longer than *Homo sapiens* have been recognizable as a species, but their reign mostly ended sixty million years ago. Where alligators persist on golf courses and other suburban settings, dogs are unfortunately among their favorite prey. For example, a twelve-foot-long problem gator on the Suwannee that was dispatched by the authorities upstream from White Springs had a dozen dog collars in its stomach. While heading our car for a canoe trip down that same stretch of river, I counted ten dogs, four with collars. You do the math.

While many people are enthralled by hawks, coyotes, and alligators, the suburban acts of predation that are of most concern are instigated by the dreaded fire ant (*Solenopsis invicta*). Fire ants, which were introduced from South America, eat anything that walks slowly, crawls, slithers, or lies down to rest. Nestlings of bobwhite quail and other ground nesters often succumb to fire ants. Unlike fat cats that tend to stay near the ground, fire ants readily climb into the treetops for those nestlings.

44. Fall in its Florida Glory

I have to admit to bridling a bit when transplanted northerners start waxing eloquent about the brilliant autumns they are missing in Massachusetts and New York. I try to get them to look around and appreciate our own annual parade of foliar magic, but they are too wrapped up in memories of maples. I could also remind them that it's not unknown in the north for Autumn to last only a few days if heavy rains or early snows knock the colorful leaves to the ground, but why be petty? Sure, the autumnal displays of reds and yellows in the North can be breathtaking, but why do they resist becoming attuned to the more subdued and prolonged seasonal changes in the South?

Unsatisfied with the "because they are beautiful" explanation for the flare up of color in the autumnal foliage of some tree species, scientists continue to explore the ecological – and hence evolutionary – advantages of leaves turning red or yellow. The reigning hypothesis derives from the observation that the red anthocyanin pigments are synthesized in anticipation of leaf drop. Anthocyanins are potent anti-oxidants that might protect the enzymes responsible for nitrogen resorption from attack by free radicals. Getting the precious nitrogen out of leaves before they fall is apparently something that deciduous trees care a great deal about. In contrast to red, the yellow carotenoid pigments just become more obvious when the green chloroplasts turn into colorless gerontoplasts and the green color of the leaves consequently diminishes.

Depending on how you reckon, autumn in Gainesville can begin as early as the first cold front in October and last well into January, with "peak" color either sometime around Thanksgiving or Christmas. Given that our flora is assembled from both species that moved up from the subtropics and species from the temperate zone that persisted here through the Pleistocene ice ages, it's no wonder that our seasons are a bit scrambled. Our red maples and white ashes have it straight, at least by northern standards. In contrast, some of the paw-paws apparently can't figure out if it's fall or spring, perhaps because the day lengths are the same in October and March.

Right about when blackgum leaves are becoming tinged with purple, narrow-leaved paw-paws flower in profusion across our pastures and sandhills. Given that these episodes of fall flowering don't ever seem to yield fruit, it must be a behavior upon which natural selection is working overtime. But coupled with the fact that I have not yet encountered a paw-paw seedling in the wild and that their giant taproots indicate their great ages, I suspect that selection against fall flowering genes is going to be a slow affair.

The three-month-long progression of autumn color in northern Florida unfortunately goes undetected by many. People may notice the red of red maples and dogwoods in November, but, unless they frequent our swamplands, they may miss the deep purples of blackgum and the burnt orange of cypress. The brilliance of the dreaded Chinese tallow tree must also be admitted, but I would be happy to see the last of that exotic invasive pest. Sweetgum trees also put on quite a show, especially the big canopy trees. Young sweetgums, in contrast, seem to try becoming evergreens, perhaps a reflection of the geographical range of this species, which reaches down to Guatemala.

Our sixteen native species of oak provide great variation on the theme of leaf fall and renewal while they demonstrate the breakdown of the evergreen-deciduous dichotomy at our latitude. At one extreme, shummard and basket oaks drop their leaves in the fall and leaf out again in the spring (in the northern manner). At the other end of the spectrum, live oaks are evergreen except for a few days in March when their crowns get quite sparse as old leaves are exchanged for new. Live oaks go on to produce another crop of young leaves in the summer, but these go unnoticed against the backdrop of green. Apparently live oaks' summer leaves are better adapted to the rigors of that season than the leaves of March. Turkey oak, and a number of its northern relatives, displays yet another twist on the leaf exchange story, one that has inspired scientists and mythmakers from the Etruscan Mountains of Europe to the Finger Lakes of New York.

Numerous oaks, beeches, and their relatives have the peculiar habit of retaining dead leaves on their branches through the winter. There has been some discussion of this topic in the scientific literature, mostly centered on the hypothesis that these so-called marcescent leaves help the leaf-retaining trees to balance their nutrient portfolios. Perhaps, but the explanation that I find more compelling reflects the wisdom of indigenous mythmakers and yarn-spinners from at least two continents.

I can't remember the details of all the marcescent leaf myths I've read, but I do recall that many start with an account of how, when the first winter came, the North Wind blew all the leaves off all the trees. When spring finally returned, the trees called a meeting during which they decided to prepare themselves for winter's return. The magnolias, pines, live oaks, hollies, and turkey oaks worked hard all summer to build leaves that could withstand the brunt of the North Wind, which they were sure would return. In contrast, the maples were convinced that winter was a once-only phenomenon, the ashes were too busy being beautiful to bother preparing themselves, and the cypress trees didn't even hear about the meeting. When winter returned, the maples and ashes and cypress trees all lost their leaves to the first puff of the North Wind, whereas the magnolias, live oaks, pines, and hollies resisted without trouble. The turkey oaks tried and tried to keep their leaves green, but, finally, they had to accept that, despite heroic efforts, they were failing. Then, as their leaves were dying, the turkey oaks decided that, in one last act of defiance, they would hold onto their brown leaves and flap them in the face of the North Wind all winter long, just to annoy him.

Another story about marcescent leaves involves a fair, young maiden who is obliged to descend into Hell when the last leaf falls in the autumn. This sad fate was the result of some shady deal between the maiden's mother and the Devil. I recall that the maiden was helped out of this dilemma by an oak tree and a bear, but I have to admit to not having those portions of the story straight. However the story really goes, those marcescent brown leaves have a beauty that deserves recognition, even if their purpose remains elusive.

Autumn, Florida style, takes us on a leisurely stroll through the color spectrum. Plus, you have time to enjoy the slow-changing kaleidoscope without fear of an early snow knocking the leaves to the ground. All of the colors are there for our enjoyment, only not all simultaneously.

45. On Georgian Hippophages and Misnamed Water Bodies: Reinstating Pithlachocho

In a bayhead swamp, two ridges back from the east shore of what (for the time being) is still called Newnans Lake, one of the very few really big slash pines was recently, first, struck by lightning, then killed by bark beetles, and, finally, toppled over. Fortunately, it wasn't the world-record-holding slash pine, which measures 135 feet tall and nearly five feet in diameter. At one time, there were probably lots of pines in that bayhead, but now there are only a dozen big ones, no small ones, and bunches of bays.

The bay trees that give bayheads their name have leaves that look like those used to flavor pasta sauce. One of those bays, swamp bay, is a close relative and a suitable substitute for the Mediterranean bay of commerce, whereas the others are only remotely related. For example, bullbay is a magnolia and loblolly bay is an overgrown camellia. Florida bays flourish in exceedingly nutrient-poor sites, where water stands above the surface much of the year. One peculiarity of our particular bayhead is that the bay trees are flourishing in the sparse shade of towering slash pines.

The fall of one of the giant pines provided me with the opportunity to read the history of this curious wetland and to reveal when the big pines were born and how they've fared since. This history was revealed after I sliced a disk of wood from the fallen tree's base, sanded it, and counted and measured the annual rings in its wood.

Judging from the narrowness of the growth rings produced over the past forty years, the fallen tree seemed to have suffered from competition with the bays growing in its shade. In contrast, from the 1920s until the 1960s, the pine grew steadily at the moderate pace of about half an inch in diameter per year. Fertilizers leaching from the adjacent citrus grove might have helped the pine cope with the otherwise nutrient-poor bayhead soil. Really rapid growth of nearly an inch per year only characterized the early life of this now dead pine. For the first twenty-five years after its birth in the early 1800s, growing conditions were apparently superb. The lack of smaller slash pines suggests that few trees of this species managed to get established since that time, which makes you wonder what led to the birth of this grove of now-giant trees.

With the appropriate technology, highly trained dendrochronologists can learn a great deal more about history than simply whether growing conditions were good or bad. Looking at the rings of my fallen slash pine in more detail, for example, indicated that the seed from which it grew germinated in 1812 or 1813. At that time, the so-called "Patriot Army" of Georgian thugs was harassing the multicultural and multiracial Seminole nation. Although the USA and Spain were supposedly at peace, in the soon-to-be-burned national capital, President James Madison and his war hawk buddies weren't acting that way. Having declared war on Britain, Madison was also trying to get Spain to cede Florida to the USA so that it wouldn't fall into the hands of the British again. While his regular troops were occupied up on the Canadian border, he was happy to have militiamen from Georgia pestering the Spanish in the South. Meanwhile, he was overseeing the reconstruction of the White House, which the Brits had burned.

I believe that the seed that grew into that big pine germinated in full sun on the ash-enriched surface that was left after Colonel Daniel Newnan torched the area during his ignominious retreat back to Georgia. By all accounts, Newnan and his band of ruffians weren't having a great time in Florida. Although they later claimed victory, they spent many days pinned down inside a tiny breastwork near the shore of what had been, and will soon again be, known as Lake Pithlachocho ("Lake of the Big Boats" in Hitichi Seminole). Ravaged by hunger, the Georgians resorted to eating the Colonel's horse. Luckily for them, the Seminoles (whose beloved eighty-year-old chief, King Payne, had been mortally wounded during one of the skirmishes) either grew bored of the standoff or were sickened by the smell of rotting horse meat. Either way, they left Newnan and his dwindling band to limp home. I suspect that, as a parting gesture, the frustrated Georgians set fire to the forest. Although this fire burned quickly through the longleaf pine uplands that surrounded the bayhead, it smoldered for weeks in the temporarily dry wetland, killing the trees, consuming the organic soil, and thus temporarily providing conditions suitable for germination of slash pine seeds.

The campaign to change the name of Newnans Lake back to Pithlachocho gained momentum during the drought of 2002-2003 when, in keeping with the lake's Seminole name, the exposed muck revealed more than one hundred dugout canoes. There were so many archaeologically important dugouts – some of which pre-dated the Seminoles by 5000 years –, that museum staff had to push some deeper into the muck to keep them from oxidizing. Whatever you think of the proposed name restoration, the bumper sticker that Hutch had printed should now make sense: "Expurgate the Georgian Hippophage: Reinstate Pithlachocho."

46. Religion in Spider Season

Damn Yankees, always complaining about how they miss the seasons in Connecticut and elsewhere up in what is called the "temperate" zone. Here in the South we even have a season unheard of in the North – spider season, in all its webby glory. Unless you are completely insensitive, the taut yellow webs of golden orb spiders stretched across every woodland trail roundabout October will give you plenty of opportunities to raise your hands in the air in celebration of the season.

Since moving to Florida, spider season has been a religious time of year for me, but the nature of that religion continues to evolve. After running into hundreds of spider webs, I soon learned to wave a stick in front of my face to clear the way. As I stick-waved, I chanted softly – it isn't really that odd for a guy alone in the woods. What was odd was that I soon recognized that my chants included odd bits of liturgical Latin recalled from Sunday morning masses back during the Eisenhower administration: *"et incarnàtus est de spìritu sancto,"* and so forth.

The thumb-sized, bright yellow banana spiders that weave the huge webs that inspired my spirituality are the females of *Nephila clavipes*. The males, in contrast, are small, numerous, and scattered around the peripheries of the web that is the handiwork of the female of their dreams. As a male *Homo sapiens,* it bothers me that, while consummating their love, the female spider eats her mate. Although the males don't spin their own webs, they do consume the leavings of their hoped-for mate. This means that they, in a sense, are just cleaning up after her and are not entirely food parasites.

My spiritual relations with banana spiders recently evolved into a more respectful one. Instead of tearing down the webs that would otherwise ensnare my face, I now, if at all possible, bow down and pass under them. The justice in that motion was confirmed when my chants somehow switched from Latin to Timucuan, a language not spoken at Flamingo Hammock for more than 200 years: *"honihe casi pono emelac ema*."

This miraculous linguistic transition makes me feel that the Timucuan held golden-orb-weaving spiders in some esteem. Perhaps they spun their bowstrings from the golden silk, as did the Yanamamo and other tribes in South America. Or perhaps, like Rima, the main character in William Henry Hudson's exquisitely romantic jungle novel *Green Mansions*, they used the silk to weave diaphanous clothing.

47. Trying To Eat Tread-Softlies

It was a mistake to walk out into the field in sandals. Since I wasn't going far, I hadn't bothered with shoes or boots. I should have known that there was a nettle out there just waiting to zap me.

We try to keep the area near the house and most of the trails free of *Cnidoscolus stimulosus*, the plant referred to as either stinging nettle or tread-softlies. Our recommended control method for this noxious native is to approach the plant warily and, be-booted, grasp its stem between the toes of your boots and lean back, thus breaking off the plant, preferably below ground. Repeated yankings are required to kill the plant because, below the rupture point, the taproot swells into a cigar-shaped tuber that plunges another foot into the ground. And, if you look closely, you'll notice that the resprouts have exaggerated densities of stinging hairs (trichomes) on their leaves, stems, and fruits. Eventually, if you are persistent, the plants succumb. Herbicides, like glyphosate, also kill stinging nettles, but I found that the big poisoned plants are often replaced by several small ones, perhaps germinating from buried, dormant seeds.

Botanically, although we too have some urticas, our nettle has no relation to the common nettle of the north, *Urtica dioica*. Northern nettles have nondescript flowers and, typically, toothed leaves, whereas ours have pretty white flowers and three-pointed leaves that exude white latex when damaged. The nettle of the north is related to hemp, fig trees, and mulberries, whereas ours is in the family that includes rubber trees, cassava, croton, and tung. Most importantly, while the stings of the northern nettle definitely hurt, our species really packs a punch.

Tread-softly stings hurt for both physical and chemical reasons. The glass daggers that cover tread-softlies are stiff, brittle, and chock full of a cocktail of compounds from which only serotonin has been isolated. Serotonin is a neurotransmitter needed for normal brain function, but the compound is also linked to depression, autism, eating disorders, schizophrenia, obsessive/compulsive disorder, premenstrual syndrome, anxiety, panic disorder, seasonal affective disorder, extreme violence, hostility and aggression, suicide, migraine, manic depression, and addiction. I'm not sure about all or any of these effects, but tread softly stings really do smart. And although Hutch brags that, as a youngster, he dueled his friends with tread-softlies (grasped by their stingless roots, I presume), sensible people give them a wide berth.

So far, this species doesn't sound like something any sane person would want to eat, but then, I've had the pleasure of eating the famous *chaya* or "tree spinach" of the Yucatan, which is prepared by cooking the young foliage of close relatives of tread-softlies (*C. acontifolius* or *C. chayamansa*). With a taste somewhat akin to our familiar spinach, *chaya* is richer in protein, fiber, minerals (calcium, potassium, and iron), and vitamins (ascorbic acid and beta carotene). Tea brewed from *chaya* is also reputed to provide therapeutic benefits for non-insulin dependent diabetic mellitus (NIDDM). All of this sounds great, but because I have intimate knowledge of both our own tread-softlies and *mala mujer* ("evil woman"), a species of *Cnidoscolus* found in Central America, I approached the culinary trials with our native species with due caution.

My first attempt at turning the tread-softly problem into a solution was to excavate and eat their roots. Armed with a long-handled spade, I ventured forth into the pasture, dug down a foot or so, and extracted several of the Tiparillo-thick tubers. At first, I tried peeling them, but then I gave up and just tossed them into boiling water. After cooking for about ten minutes, the hard-won result was a boiled tuber that had a delicious, almost nutty flavor, much tastier than an "Irish" potato or cassava.

But digging is too much work and, anyway, my principal focus was on tread-softly foliage as food. Gloved and in long sleeves, my son Antonio and I collected a pot full of fresh plants and put them on the stove to boil. Our approach was to cook the plants for different lengths of time, removing a few leaves every fifteen minutes. Since the idea of piercing the roofs of our mouths with toxin-packed glass syringes appealed to neither of us, we settled for using a hand lens to check for intact trichomes on the cooked plants. After five minutes of boiling, the stinging hairs still looked lethal. An additional fifteen minutes of cooking seemed to have little effect. After forty-five minutes of a rolling boil, the needles were still terrifying and we finally tossed the leaves out, unsavored.

Perhaps we were cowards. Perhaps boiling had disarmed the still dangerous-looking trichomes. In our defense, they were still brittle when finger prodded. In any event, we decided to cease in our pursuit of eating tread-softly foliage, and we left this to other, more adventuresome investigators and gopher tortoises, which eat them with relish. Instead, we'll soon attempt to make food from the real Mayan chaya now growing in our garden.

48. Mockernuts

*"We rely on your justice and humanity;
we hope you will not send us south, to a country where neither the hickory
nut, the acorn, nor the persimmon grows"*

Neamathla, Mikasuki hard-liner, 1824,
during negotiations for the Treaty of Moultrie Creek

Mockernut (*Carya tomentosa*) is one of six native Floridian species of hickory. Its geographical range is from the shores of Lake Erie down to just south of Ocala. Pignut hickory (*C. glabra*) shares that geography, but its nuts were not nearly as important to pre-Columbian Floridians. Mockernut can be distinguished from pignut by its fuzzy (tomentose) leaves, thick twigs, and brilliant ochre autumn coloration. In 1539, near what we now know as Gainesville, Hernando de Soto described a large grove of what he took to be walnut but was most likely mockernut. Sadly, victims of over-logging and fire suppression in the woodlands in which they once abounded, mockernut trees are now scarce.

Mockernut's geography was relevant to the Seminoles, and the Timucuans before them, because its nuts were an important food and source of vegetable oil. As a member of the Juglandaceae – the family that features both pecans and walnuts – perhaps the exquisite taste of mockernuts comes as no surprise. What does surprise me is how hard it is to get food from a mockernut – it's a very hard nut to crack whose deeply divided seed is then hard to extract.

After our collection of nutcrackers proved worthless when faced with a mockernut, I rolled out the live oak pestle with ten-pound (pignut) hickory mortar I'd made for my wife. I mention that lovingly crafted device with some chagrin because, for reasons that still elude me, she has never used it. Unfortunately, when I tried to use it myself on mockernuts, I managed only to make mockernut butter with a liberal mixture of chipped shells.

William Bartram described how the Seminoles he visited in the 1770s made hickory soup: "They pound them to pieces, and then cast them into boiling water, which, after passing through fine strainers, preserves the most oily part of the liquid; this they call by a name which signifies Hiccory milk *[sic]*; it is sweet and rich as fresh cream, and is an ingredient in most of their cookery, especially homony and corn cakes." Basically the same processing instructions can be found in numerous books about eating off the land, but I wonder whether the authors of those books ever tested Bartram's method. I know that, despite having tried it (and every variation on the theme I could devise), I have not yet been able to make hickory nut soup or to extract hickory oil. Instead, I end up with a bitter, grey-black mass laced with shell fragments without any indication of oil separation. Perhaps the common name of *mocker*nut was coined to capture just this sort of frustration. In any case, I offer three explanations for my failure:

1. The Seminole men who told Bartram how to make hickory nut soup were limited in their culinary knowledge.
2. To maintain their intellectual property rights, the men withheld some critical steps in the process.
3. I am culinarily challenged

There are hints in the ecological literature that, prior to the plowing, fencing, and fire suppressing of the past century, an ecosystem dominated by longleaf pine (*Pinus palustris*), southern red oak (*Quercus falcata*), and mockernut was once widespread across the South. Few people are aware of this ecosystem type because it was the first to be cleared by colonizing farmers, since they too appreciated its relatively nutrient-rich soils. More recently, fire suppression continues to take its toll and these rich soils have become dominated by thin-barked oaks and other fire-sensitive species.

Judging from the scattered southern red oaks at Flamingo Hammock, the even scarcer mockernuts, the occasional longleaf pine, and the presence of some subsoil clay, I now recognize that much of our uplands probably once supported this now-rare ecosystem type. Logging, hog rooting, cattle grazing, and decades of fire suppression easily account for the scarcity of longleaf pines, and our southern red oaks are dying at an alarming rate for unknown reasons, but the cause of our mockernut deficit was not initially apparent.

Then Hutch, while he was searching through land records at the Alachua County Courthouse, found a record from the 1950s of a sale of hickory timber by a former owner of our property. Presumably, the hickory in question was mockernut, the wood of which was used for the shafts of my first set of golf clubs in addition to making tool handles, tennis rackets, and for smoking hams. Now we are left mostly with stump sprouts and wolf trees of a species that was once presumably more common.

Former prominence of mockernuts in our landscape is also suggested by archaeological evidence from a nearby 1,000-year-old village site. The excavation, executed by the Florida Museum of Natural History, revealed the usual assortment of fish bones, alligator scutes, and pottery shards, but it also uncovered several large subterranean food-storage bins. Based on the abundance of nut shells, the archaeologists concluded that the containers were used for mockernut storage. This revelation, coupled with my frustrated attempts at harvesting mockernuts, caused me to revise my vision of pre-Columbian landscapes and has otherwise wreaked havoc with my concept of "natural" Florida.

I now believe that Seminoles, Timucuans, and earlier inhabitants managed substantial portions of our land as a mockernut orchard. To facilitate harvesting, they burned the understory in their orchards just before nut drop in October. I recognize that October does not fall in Florida's "natural" fire season, but I challenge you to find fallen mockernuts without having first burned off the underbrush. Based on the results of another experiment in which I carefully planted 100 nuts (only to have them all harvested by gray squirrels within a week), I also propose that our predecessors ate a lot of Brunswick stew. Further support for this conclusion is that a good place to collect mockernuts is near the house of a pot-hunting neighbor with a squirrel vendetta.

In the absence of sufficient grey squirrel control in the area I am trying to restore to mockernut prominence, I have taken to outplanting seedlings that I raise in tall pots in a protected area. After six years in the ground, many of the seedlings, though still alive, were only knee high. Even the ones that I accidentally mowed or burned off were about the same height. Some shovel work revealed that, like longleaf pines, mockernuts were investing heavily belowground; those knee-high seedlings were supported by taproots deeper than I was willing to dig. Now, after an additional six years in the ground, some of those seedlings are growing taller and fatter at a remarkable rate. Apparently, they were waiting until they had a deep enough root system – and perhaps enough carbohydrates stored belowground – to fuel rapid growth to a size at which they will no longer be fire susceptible. Those that do pass the gauntlet of early challenges may live up to 500 years.

The taste of mockernuts, reinforced by these rather iconoclastic insights about our land's history, caused me to radically rethink our land management practices and restoration goals. Instead of being pristine, virgin, unsullied, or otherwise "natural" until being despoiled by Europeans, I now believe that, for at least a few thousand years, much of our landscape was humanized, domesticated, and otherwise managed. Certainly, North-Central Florida was not one big mockernut orchard, but it was also no wilderness area where the hand of man had never set foot.

POSTSCRIPT/ADDENDUM

As this book went to press I came to realize that, to the list of three explanations for my failure to make mockernuts into food, I now need to add a fourth and perhaps a fifth: inadequate scholarship and sloth. My oversight of an important article on the making of hickory nut soup was recently corrected by an archaeological colleague. In their 2001 article entitled "Ethnobotany of *ku-nu-che*: Cherokeee hickory nut soup," authors G.J. Fritz, V. Drywater Whitekiller, and J. W. McIntosh clarify that I stopped my nut processing too soon. The traditional approach, which is still employed in parts of Tennessee, involves pounding the nuts and then forming the mixture of shell fragments and nutmeat into softball-sized masses. The balls can be stored frozen for months and, when it's soup time, dissolved in boiling water, strained to remove the shell fragments, sweetened or salted, and eaten with rice or hominy. *Ku-nu-che* soup reportedly remains a much-relished holiday treat in Cherokee communities. The authors also point out that the caloric and protein contents of hickory nuts are far higher than other nuts, which reinforces the argument that they were a critical food source. Even more intriguing is their suggestion that mockernut orchards were prime sites for domestication of sumpweed (*Iva annua*) and goosefoot (*Chenopodium berlandiera*), both highly nutritious seed crops that deserve more attention from modern farmers.

49. Fox Squirrel Aspirations

I aspire to fox squirrels. I long for the day when I can sit on our deck and watch big, bushy-tailed fox squirrels cavorting about in the pine savanna we've restored in an abandoned pasture. On that day, I will have reached the restorationists' nirvana. But being botanically inclined and recognizing that we have a long way to go before fox squirrels will once again be comfortable at Flamingo Hammock, I'll continue to celebrate every blazing star and smile at gopher tortoises, but fox squirrels will remain my ultimate objective.

If you've ever seen a fox squirrel, perhaps my aspirations make sense. That photogenic species should be featured in campaigns to restore and protect pine savannas in the South. At twice the size of grey squirrels, otherwise known as brush-tailed tree rats, a mature fox squirrel is a substantial beast. In addition to being cute and cuddly looking, every fox squirrel has a different combination of coat colors. Genetically, fox squirrels are mosaics. While all individuals in the subspecies in our region have a black head with a white nose and ears, the rest of their bodies are patchworks of tan, brown, orange, white, and yellow, often with the striped or grizzled hairs referred to as agouti. Given this variation in their pelage, not only can individual beasts be easily differentiated, but photographers and artists won't soon run out of material.

The evolutionary significance of the peculiar pelage variation of fox squirrels attracted the attention of Richard Kiltie, a researcher in Gainesville. He wondered whether fox squirrels benefit in any way from having dark heads and multi-colored bodies and tails. Given that, from the perspective of raptors, bobcats, or other predators, a fox squirrel is just a nice package of fresh meat, Richard wondered whether these tasty prey items escape predation by being cryptic. Since they spend a lot of time on the ground foraging for mushrooms and fallen seeds in habitats that are charred by fires, he thought black coats might've been favored. A survey of skins from natural history museums around the South revealed some consistency with this hypothesis, but no clear patterns emerged, so Richard carried out a series of predation experiments using fox squirrel mimics of various hues.

The results of Richard's elegant experiments can be read on the pages of the *Biological Journal of the Linnean Society*, so I won't bother with all the technical details here. But I should say that the study involved a clothes line, string, indoor-outdoor carpet, paint, fake fur, and (temporarily) captive red-tailed hawks that were trained to attack fox squirrel mimics of different colors as they were pulled across different backgrounds (also of different colors) in his backyard. Using the time it took the hawks to start moving towards the moving or stationary fox squirrel models (to which dead mice were attached) as his response variable, Richard found that there are indeed benefits to being different. His hawks responded more slowly to moving fox squirrel mimics that were patchworks of colors than to those of a single hue.

Fox squirrels are the "flagship" species of pine savannas. Although the South Florida subspecies frequents cypress and even mangrove swamps, ours depends on large tracts of open-grown longleaf pine trees. With a few large, hollow, dead standing trees scattered across an area of at least forty acres, you have the perfect fox squirrel habitat. And if a pine savanna has a viable population of fox squirrels, then you should look for red cockaded woodpeckers next, and you are also almost certain to find gopher tortoises.

Our fox squirrels, *Sciurus niger shermani,* are discriminating longleaf pine specialists. While they'll eat plums, acorns, mushrooms, and insects, they really love longleaf pine seeds. One limitation on this specialization is that, while longleaf seeds are large and quite tasty, good cone crops are only produced at five to seven year intervals. This habit of masting is shared by live oaks and turkey oaks; bumper crops of acorns are only occasional. It's thought that, between mast years, populations of seed predators decline due to lack of food. Then, when entire populations of trees all reproduce at the same time, there aren't enough seed eaters to consume the entire crop and a new cohort of seedlings is thus assured. This on-again-off-again behavior might be good for live oaks and longleafs, but it makes for some lean years for fox squirrels.

Although you can still occasionally glimpse fox squirrels along roadsides, populations of these beautiful beasts are diminishing to the point that the Florida Fish and Wildlife Conservation Commission now considers them a "Species of Special Concern." Hunters take a few, but the biggest problem fox squirrels face is habitat loss due to development, pine plantation establishment, and fire suppression in the patches of pine savanna that do remain intact.

There's a fox squirrel translocation program in our area, so if you have forty-plus acres of intact pine savanna, you might think about providing a home for animals threatened by suburban sprawl. I hope that Flamingo Hammock will someday qualify. As that day approaches, we'll train the dogs to leave the fox squirrels alone, perhaps by rigging up some of those fox squirrel models with some deterrent. The dogs already know to act as if gopher tortoises are invisible, but big bushy-tailed, multi-hued squirrels running along the ground might represent a bigger temptation.

50. Ancient Palmettos

A few weeks had passed after a fuel-reduction-and-mockernut-collection facilitation burn, and I was curious about what was sprouting. Green up after late-season burns is not as rapid as during the summer, but the saw palmettos surrounding one of the little sinkholes were gaily pushing up bright, new leaves. I remember that the fire was very intense at this spot during the burn. It had ripped through the palmetto leaves, which are justifiably referred to as "green gasoline." Watching the fire crown-out in the overtopping laurel oaks had also given me great pleasure; hopefully, the oaks will at least be top-killed. I wasn't concerned about the survival of the ground-hugging palmettos, since they are virtually impossible to kill with fire.

With the fronds burned off and the creeping palmetto stems revealed, a particularly long one caught my eye – it had to be nearly eight feet long, which got me wondering about its age. I know from watching them that saw palmettos produce an average of 3.2 leaves per year, perhaps one or two more than that during the year after a fire. The leaves are crammed together on the stem, but, by carefully pulling apart some dead palmettos and extrapolating a bit, I estimated that, on a linear foot of stem, there are 47.3 leaves. Continuing on with these calculations led me to conclude that this particular saw palmetto was born in July 1625 (give or take a century). This plant, this lowly and seemingly insignificant palmetto, may very well be the oldest living thing in the area.

Careful scrutiny of the location of this nearly 400-year-old saw palmetto revealed that it most likely germinated though the scat of a young, male brown bear who was suffering the consequences of eating too many palm fruits. He stopped only briefly because he caught a whiff of smoke, which suggested the proximity of a human village from which he wanted to distance himself. The young saw palmetto plant was likely munched on by cows and horses that were trying to find something to eat in, what was for them, a New World. Its first flowers were most likely pollinated by native bumblebees, at least if European honeybees were still scarce in the area.

Over the centuries, saw palmetto fruits were eaten by other bears as well as coons, turkeys, foxes, tortoises, and humans with taste buds more refined than mine. This now-old plant watched the demise of the Timucuans and the arrival and then departure of the Seminoles, if that is the appropriate way to portray this cultural transition. As a mature member of its ecological community, it somehow survived the eradication efforts of the cracker settlers who row-cropped on the other side of the creek, close to the ancient Timucuan village of which they were probably unaware. Now the palmetto is once again safe, venerated (at least by me), and protected with my ax from the encroachment of oaks and others that might cause it stress.

While still on the topic of saw palmetto longevity, I should admit the limitations of low-tech approach to scientific research. Researchers armed with DNA sequencers and sophisticated mathematical models have recently estimated that large saw palmetto clones might be up to 10,000 years old. That means that, early in their lives, some of our saw palmettos contended with mastodons and giant ground sloths as well as bears and fires. Increasing interest in growing drought-tolerant native species coupled with the rediscovery of the health benefits of eating palmetto fruits are elevating saw palmettos in the esteem of many people. In addition to various reputed restorative properties, saw palmetto fruit extract is widely recommended as an herbal treatment for benign prostatic enlargement (even if clinical trials do not quite support this claim). Despite these qualities, countless thousands, some even older than the plant by our sinkhole, are torn out of the ground each year during pine plantation establishment, suburban development, and pasture management.

The scarcity of saw palmettos at Flamingo Hammock stimulated me to pull a dozen wrenched palmetto stems from a wind-rowed pile in a slash pine plantation and transplant them around the property. This effort amused Hutch, who pointed out that some cracker farmer had probably dedicated much of his life to eradicating saw palmettos, only to have me replant them. But I believed in what I was doing, and so I did it with diligence. I watered the transplants weekly for several months, but, at that point, I was sure that they were all already dead. Sure enough, when I pulled them up (which wasn't hard), not a single new root was to be seen.

From the ashes of my transplanting debacle, a research project focused on the issue of why it is so hard to transplant saw palmettos emerged. For this effort, I wisely enlisted the assistance of then-graduate student Michelle Pinard, now a professor at the University of Aberdeen in Scotland and far from palmetto country. As a first step, we obtained permission to excavate some saw palmettos in an area slated for suburban conversion. We thus ventured forth, shovels in hand, only to find out that saw palmettos are pretty easy to dig up, unless of course you try to excavate their roots, which is a waste of time on several counts.

Saw palmetto roots should fascinate even people who aren't usually interested in root biology. Like other palm body parts, palm roots cannot grow in diameter and none of them grow thicker than a pencil. While not thick, they penetrate the ground to great depths and spread horizontally further than would seem necessary. At least partially because their roots are well endowed with air-filled tissues, saw palmettos persist in seasonally flooded soils where oxygen availability to roots is limited. Survival in wetlands is most likely also assisted by oxygen that diffuses into the plant through the numerous small, upward-growing roots that look like white-striped pipe cleaners. Remarkably, saw palmetto roots are unique in their ability to puncture the buried hardpans that perch above the water table in the flatwoods where they abound. The channels opened by saw palmetto roots later provide pathways for the roots of other species seeking a way down through the cemented layer.

Long and deep and tough as they are, if cut even several feet away from the stem, saw palmetto roots die all the way back to their origin on the prostrate stem. This peculiar response means that, when you are excavating palms, you should cut them off near the stem, which greatly facilitates transplantation.

For experimental material, we went to an area slated for development and excavated a circular clump of saw palmettos that were about fifteen feet across. This species is one of few palm species with stems that branch, so we weren't sure whether the clump represented one branched individual or several different plants growing together. Our shovel work revealed that the latter was the case: the clump was comprised of seven long-stemmed plants growing out like the spokes of a wheel. At the center of the clump, we found the saxophone-shaped stem of each plant – the result of the shoot being pulled down into the ground by retractile roots. I suspect that all of the stems in the clump germinated from the scat of a bear, perhaps a relative of the one who made his deposit at Flamingo Hammock. Each of the stems was indeed branched, but the branches only produced small coronas of leaves on the margins of the main stem.

Our saw palmetto transplanting experiment was pretty simple. Given that transplanting success with cabbage palms increases with plant size, we selected saw palmetto stems of three different lengths. We planted several individuals of each of the three sizes under shade cloth and watered them every other day for six months. We took another set of plants, cut off all their leaves and roots, and put them on a greenhouse bench, where they were misted several times per day.

Just as we predicted, the small plants died rapidly, both on the mist bench and in the ground. In contrast, most of the larger individuals planted outdoors were still alive after six months – not flourishing, but alive. What surprised us was that many of the larger plants on the mist bench were also alive, although none had grown new roots over the six month observation period. After making this observation, we excavated the transplants that had survived in the ground to check on their rooting status. While a few new roots had emerged from the large stems, none were more than a few inches long. That these rootless and nearly-rootless plants survived so long surprised us, but our results also suggested that saw palmettos absorbed water through their stems. The general lesson from this study was that saw palmettos can indeed be transplanted, but that success will require considerable patience as the roots slowly regrow.

Since completing that experiment, I've successfully transplanted several large saw palmettos, of which I am immensely proud. The fruits do indeed taste "like rotten cheese steeped in tobacco juice," as described by the shipwrecked Jonathan Dickinson, but they make a potent spring tonic. It's a lot of work to get saw palmetto re-established, but, by saving such old plants, it somehow feels a little bit like saving ourselves.

51. Onset of the Homogeocene, Continuing Extinctions, and Growing Boredom

When it comes to ecosystems, change is the only constant. We're concerned that the seas around us are rising, but, only ten million years ago, Florida was a chain of islands in a warm Miocene Sea. We worry about the human modification of our landscapes, but some of those islands developed novel ecosystems of which there are no modern analogues. As the land rose and the sea rose and fell and rose again, the Florida Peninsula became as we know it, complete with its own flora and fauna uniquely assembled from species borrowed from the north and south, plus some that evolved right here. Although this uniqueness is now threatened by widely heralded extinctions, it's put even more at risk by the homogenization of ecosystems and cultures that is occurring at an unprecedented rate over much of our small planet.

With about six hundred people moving to our fair state every day from every direction on the compass added to the millions of plants and animals being legally imported from all over the world every year, it's no surprise that Florida natives – both animal and vegetable – are being overrun. Its latitudinal span and the staggering rate at which people, plants, and animals arrive render Florida especially prone to the impacts of global swarming. I'm not sufficiently trained in anthropology to comment on the parallel processes of cultural homogenization that has half of the world's languages on the brink of extinction, but I do have a few things to say about our species' impacts on natural environments, impacts that are matched only by the effects of the giant asteroid that terminated the Paleozoic.

That so many species are introduced so frequently over so much of our landscape is central to our exotics problem, but the capacity of introduced plant and animal species to invade Florida is increased by the disruption of natural processes that might otherwise have kept them at bay. In particular, if fires still burned across the landscape every few years, if our swamps were still swampy and not drained for housing estates, and if our lakes weren't loaded with lawn fertilizer, it would be harder for many of these species to gain footholds. But, even if our ecosystems were still otherwise intact, cogongrass from Asia would still be able to invade our savannas, tallow trees would still crop up in our marshes, ardisia would still colonize forest understories, and hydrilla would still find some waterways in which to flourish.

Managers of our beleaguered natural areas in Florida spend increasingly large portions of their already-meager budgets controlling exotic invasives. These heroic efforts can be justified in many ways, but one of the most commonly invoked reasons for battling against Japanese climbing ferns, African tilapia, and South American water hyacinths is that they threaten native species with extinction. Fears of extinction are completely justified, but I am equally concerned about the boredom that will ensue long before many of our native species have completely disappeared from the face of the earth. As with loss of linguistic and cultural diversity, the result of unchecked biological invasions will be a numbingly homogenized world.

Prior to jet travel, ornamental horticulture, and the human population explosion, transitions between geological eons, eras, periods, and epochs were heralded by events such as widespread glaciation, asteroid impacts, and supercontinents rending apart. Only one species, *Homo sapiens*, has had the dubious distinction of having wrought so much change on the face of the Earth so as to warrant my naming of this new epoch the Homogeocene. The etymological root of my preferred term can be traced back to the name of our own genus – which is fully warranted given that we are the principal perpetrators – as well as to the word "homogenize," both of which come from the Greek *homos,* meaning "one and the same." Although, in the literature, my Homogeocene lost out to the Anthropocene, it's the same idea.

Since humans require very few species of plants and animals for survival, our species will most likely survive the Homogeocene. Sometimes, when struggling with the Spanish subjunctive, I think it would be better if the Tower of Babel finally crashed and we all spoke Mandarin. We could also buy all we need to survive at Wal-Mart. And McDonalds could feed us for the duration of our short lives. Do we really need all those works of art that can't be distinguished from industrial accidents or all those poems that make little sense to anyone but the anguished poet? And if there really was more than one disco song, do we need them all? Finally, and closer to home here in Florida, do we really need six species of pines, fifteen of oak, and I-don't-know-how-many of grass?

If you objected to any of these assertions, or even feel fondness for species that you haven't yet learned to differentiate, then I share with you this appreciation of diversity. I also like to travel to new places to see new things. I am therefore as offended by familiar chain stores in otherwise exotic places as I am by exotic species in familiar places. Just as I don't want to see American pines when I go to Africa, I don't want African fish taking over our lakes in Florida. Someday, I'd like to visit the species-rich hardwood forests of Japan and see ardisia fruiting in the understory, but I don't like seeing that same ardisia choking out natives here in our state parks. I accept that most of what I eat is not native to this region, and I understand that change is inevitable, but when the impending boredom of the Homogeocene looms large, I find myself compelled to round up African air potatoes, to yank out Asian tradescantia, and to otherwise keep my little bit of Florida, well, Florida.

52. Final Home Found

As I've written about many of the natural-history marvels one can encounter daily here in the sandy lands of the South, I've mentioned that, over the years, I've occasionally hesitated to accept the local. If he heard about this hesitation, my late father would undoubtedly have said something along these lines: "Jezz, yer from Jurzee for chrissake." Meanwhile, my highborn wife (nearly 12,000 feet in the Andes) accuses me of being a wannabe redneck, a criticism with just an element of truth. I always did like Jed Clampett, and I have, more than once, unloaded pickups by driving backwards rapidly and stomping on the brakes, but I only pass Jeff Foxworthy's test of redneckedness if the scores are curved. Instead, I credit the light shade of pink on my neck for the opportunities I've had to breach the town-gown divide, which is sometimes unfortunately wide and deep in Gainesville. That I bought land and started building a house only two years after I arrived might indicate that whatever initial hesitation I felt about becoming a Southerner or a Floridian were more than compensated for by my rapidly growing sense of the place.

The final indication that I found home in Florida is that I plan to be planted here when my time comes. Like most everyone else, I am not planning on an early demise, but upon that eventual event, I have made arrangements for my earthly remains to rest in the earth at the Prairie Creek Green Cemetery. My choice of a green burial reflects my dedication to environmentalism, a cause I want to continue to serve in the afterlife. Basically, I want the environmental damage that I cause to cease when I die. The cremation alternative would add far too much to my already-large carbon footprint, not to mention the mercury that would be volatilized upon heating my body 2,000 degrees. Being embalmed with nasty carcinogens also holds little appeal. So, wrap me in a tie-dyed cotton shroud and bury me under one of the live oak trees that I liberated.

Prairie Creek Cemetery is well trodden, if not hallowed, ground. Before Hutch managed another of his magnificent conservation land deals that included that property, it was part of a game ranch run by the scion of a wealthy family, who later ran that fortune into the ground. He had surrounded his 700 acres with a ten-foot-high fence and stocked it with big game animals that included Watusi cattle and Asian black bucks. The operation was right out of a Carl Hiaasen novel, with a bit of Harry Crews thrown in for good measure. When Alachua Conservation Trust (ACT) purchased the land, the exotic game animals were all gone, except for a male single-horned Barbary goat that, years later, is still occasionally seen.

ACT's land stretches from Prairie Creek east to County Road 234, which runs from Windsor through Rochelle, crosses the Gainesville-Hawthorne Rail Trail at Witness Tree Junction, and continues on to Micanopy. For hundreds of years before the trains stopped running several times a day from Gainesville to Palatka, the CR 234 corridor was part of the route from Fernandina to King Payne's Town. Along that route, from the late 1700s through the 1800s, live oakers commissioned by shipwrights from as far away as Massachusetts combed the woods for appropriately-shaped live oak branches, crotches, and crooks for the construction of schooners, frigates, sloops, and ships of the line. Even before that, as part of the Arredondo Land Grant, the area was heavily grazed by Spanish "cracker" cattle and most likely plowed and planted with cotton, cane, and corn. Then, there was the week in October 1812 that Colonel Daniel Newnan and his rag-tag band of Georgian slave raiders spent, pinned down and without provisions, behind a crude bulwark after they met up with a group of armed Seminoles.

The live oaks that shade parts of the Prairie Creek Cemetery may not be large enough for membership in the Live Oak Society, but they are nevertheless impressive. It was the parents of these trees that supplied shipwrights with the durable sternposts; transoms; bow timbers; futtocks; aprons; compass pieces; mortar beds; and lodge, thrust, and hanging knees. The wagons that hauled that live oak wood most likely had hubs, axles, and felloes constructed from the same. The live oakers were probably poachers who had moved far from the coasts, more likely due to exhaustion of the live oak resource than to constraints by the authorities. By 1799, their depredations were substantial enough for the federal government to proclaim the first national timber reserve in the USA.

If we assume that the live oaks at Prairie Creek Cemetery sprouted about 100 years ago, their short statures and spreading crowns indicate that they spent most of their lives growing in the open. Before they shaded the exotic game animals, those trees likely provided comfort to many generations of cracker cattle. It was only over the past few decades that both fire frequencies and stocking densities were too low to keep the lesser hardwoods at bay. Now, with their crowns menaced mostly by laurel and water oaks – plus some pignut hickories and sweetgums – the live oaks need human assistance if they are to continue to provide me shade. Live oaks are so susceptible to encroachment because the dense wood so prized by shipwrights also keeps their branches from swaying and beating up the interlopers when the breezes blow. In other words, the crown shyness mechanism is inoperable in stiff-branched trees.

I hope that the stewards of the land where I expect to be interred are careful about how they keep the live oak crowns free to grow. Although I advocate very selective application of herbicide to the stumps of the felled encroachers, I recognize the risk of killing the trees that the managers are trying to save. While it is clear that live oaks are very sensitive to herbicides, the transfer mechanism is not so clear. I doubt that the small quantities of herbicide use are volatilized, and I don't think that live oaks form root grafts with water or laurel oaks, but I would still recommend gradual release treatments, a few encroachers at a time.

As more of the landscape is restored to live oak and longleaf pine savannas, I hope that fox squirrels return. I'd like to picture those multi-hued and bushy-tailed beasts cavorting around, munching mushrooms and just being beautiful while I rest in peace in my final home in Florida.